INTRODUCTION TO CELESTIAL MECHANICS

ASTROPHYSICS AND
SPACE SCIENCE LIBRARY

A SERIES OF BOOKS ON THE RECENT DEVELOPMENTS

OF SPACE SCIENCE AND OF GENERAL GEOPHYSICS AND ASTROPHYSICS

PUBLISHED IN CONNECTION WITH THE JOURNAL

SPACE SCIENCE REVIEWS

VOLUME 7

JEAN KOVALEVSKY

INTRODUCTION TO
CELESTIAL MECHANICS

Translated by Express Translation Service

SPRINGER-VERLAG NEW YORK INC. / NEW YORK

D. REIDEL PUBLISHING COMPANY / DORDRECHT-HOLLAND

INTRODUCTION À LA MÉCANIQUE CÉLESTE
Librairie Armand Colin, Paris, 1963

SOLE-DISTRIBUTORS FOR NORTH AND SOUTH AMERICA
SPRINGER-VERLAG NEW YORK INC. / NEW YORK

Library of Congress Catalog Card Number 67-23413

ISBN-13: 978-94-011-7550-0   e-ISBN-13: 978-94-011-7548-7
DIO: 10.1007/978-94-011-7548-7

© 1967. D. Reidel Publishing Company, Dordrecht, Holland
Softcover reprint of the hardcover 1st edition 1967

FOREWORD

After the launching of the first artificial satellites preceding interplanetary vehicles, celestial mechanics is no longer a science of interest confined to a small group of astronomers and mathematicians; it becomes a special engineering technique.

I have tried to set this book in this new perspective, by severely limiting the choice of examples from classical celestial mechanics and by retaining only those useful in calculating the trajectory of a body in space.

The main chapter in this book is the fifth, where a detailed solution is given of the problem of motion of an artificial satellite in the Earth's gravitational field, using the methods of Von Zeipel and of Brouwer. It is shown how Lagrange's equations can be applied to this problem.

The first four chapters contain proofs of the main results useful for these two methods: the elliptical solution of the two-body problem and the basic algebra of celestial mechanics; some theorems of analytical mechanics; the Delaunay variables and the Lagrangian equations of variation of elements; the expansion of the disturbing function and the Bessel functions necessary for this expansion.

The last two chapters are more descriptive in character. In them I have summarized briefly some of the classical theories of celestial mechanics, and have tried to show their distinctive characteristics without going into details.

Finally, at the end of the seventh chapter and returning to more technical matters, I have described a method for a numerical integration of differential equations, which allows one to resolve the problem of any trajectory whose analytical solution is not required.

I should like to thank M. Jean-Jacques Levallois of the Institut Géographique National and M. Bruno Morando of the Faculté des Sciences de Paris, both for reading the manuscript and for valuable discussions.

JEAN KOVALEVSKY

TABLE OF CONTENTS

CHAPTER I

THE PRINCIPLES OF CELESTIAL MECHANICS

Celestial mechanics is essentially an application of the universal laws of mechanics to the study of motion and equilibria of celestial bodies subjected to gravitational forces. The principles of celestial mechanics are thus those of general mechanics to which we must add the universal law of gravitation.

1. The Fundamental Law of General Mechanics

We shall recall this law here because of its importance in celestial mechanics; the reader who wishes to go further into the matter is referred to any book on classical mechanics. We admit the concept of mass, as well as the existence of a spatial frame (called a Galilean or inertial reference system) in which the laws are valid. Finally, we admit a variable called time, which is experienced everywhere simultaneously (absolute time). Under these conditions, the fundamental law of mechanics is as follows:

When a material system is at rest or in motion in an inertial frame of reference, the set of vectors $\mathbf{F}_i$ representing external influences on the system is equivalent to the set of vectors $m_i \gamma_i$ (where m_i is the mass and γ_i the acceleration of mass i) representing the inertial forces:

$$\Sigma \mathbf{F}_i = \Sigma m_i \gamma_i \tag{1}$$

In particular, if the system is a single particle and the vector $\mathbf{F}$ represents the resultant of the applied forces, then, for mass m and acceleration γ:

$$\mathbf{F} = m\gamma$$

Another special case is when no external force acts on the particle. The acceleration is zero and the motion of the particle is rectilinear and uniform. It immediately follows that two inertial systems are either fixed, or are in uniform rectilinear motion with respect to one another.

In general, we shall write our equations in an inertial system. However, in studying motion in another, different, system, we must add to the external forces, complementary inertial forces arising from the addition of accelerations.

2. General Theorems of Mechanics

These are the consequences of the fundamental law of mechanics. They present mechanics in a practical form for its application to the treatment of special cases.

A. THEOREM OF THE MOTION OF THE CENTRE OF GRAVITY

The centre of gravity of a material system moves as if the whole mass of the system were concentrated there, and as if all external forces were acting upon it.

If O is a reference point and G is the centre of gravity, then:

$$\mathbf{OG} = \frac{\Sigma m \mathbf{OP}}{\Sigma m},$$

where the mass m is situated at point P.

Putting M, the mass of the system, equal to Σm, we obtain

$$M\mathbf{OG} = \Sigma m \mathbf{OP}$$

and, differentiating twice, we arrive at

$$M\mathbf{\gamma_g} = \Sigma m \frac{d^2\mathbf{OP}}{dt^2} = \Sigma \mathbf{F_E} \qquad (2)$$

in which the second equality is formula (1), the fundamental equation.

B. THEOREM OF ANGULAR MOMENTUM

The time derivative of the angular momentum about a point O of a material system, is at any instant equal to the resultant moment of external forces about the same point O.

By definition, the angular momentum is

$$\mathbf{\sigma} = \Sigma(\mathbf{OP} \wedge m\mathbf{V_P})$$

whence

$$\frac{d\mathbf{\sigma}}{dt} = \Sigma\left(\frac{d\mathbf{OP}}{dt} \wedge m\mathbf{V_P}\right) + \Sigma\left(\mathbf{OP} \wedge m\frac{d\mathbf{V_P}}{dt}\right)$$

The first vector product is zero, and we have

$$\frac{d\mathbf{\sigma}}{dt} = \Sigma(m\mathbf{OP} \wedge \mathbf{\gamma_P}) \qquad (3)$$

An immediate consequence of this theorem is that if $\mathbf{\sigma} = 0$ (zero or central force), the motion takes place in a plane.

C. THEOREM OF KINETIC ENERGY

The change in the kinetic energy of a material system during a time interval $t_0 - t_1$ is equal to the sum of the work done on the system by internal and external forces applied during the same period.

The kinetic energy of a system is, by definition $\Sigma m V_p^2$, summed over all the components P of the system.

If $\mathbf{F_I}$ and $\mathbf{F_E}$ are respectively the resultants of the internal and external forces applied on P, the theorem is written as:

$$\tfrac{1}{2}\Sigma m V_P^2(t_1) - \tfrac{1}{2}\Sigma m V_P^2(t_0) = \Sigma \int_{t_0}^{t_1} (\mathbf{F_I}(t) + \mathbf{F_E}(t))\, \mathbf{V_P}(t)\, dt \qquad (4)$$

where $V_P(t)$ indicates that we are dealing with the velocity V_P at an instant t.

Each of the formulae (2), (3), and (4) is a consequence of (1), which in fact contains all the dynamic properties of the system. We shall nevertheless use these formulae in certain simple cases to shorten some proofs.

3. Newton's Law

Newton's universal law of gravitation specifies the nature of the forces acting on a material system, and study of the consequences of this law within the framework of general mechanics constitutes the field of celestial mechanics. The law states that *two particles A and B of masses m and m' attract each other along the line AB with a force directly proportional to the product of their masses and inversely proportional to the square of their distance.*

If $AB = r$, the force on B is directed from B to A and has the magnitude $k(mm'/r^2)$ where k is a proportionality constant called the universal gravitational constant. In the C.G.S. system, its experimentally determined value is $6.670 \pm 0.005 \times 10^{-8}$ dyne cm^2/g^2.

In section 15 we shall give the value of k in the system of units used by astronomers. Vectorially, Newton's law is written as follows:

Attractive force on B by A : $\mathbf{F_B} = kmm'\mathbf{BA}/r^3$;

Attractive force on A by B : $\mathbf{F_A} = kmm'\mathbf{AB}/r^3$.

NOTE:

Newton's law implies the instantaneous transmission of force through space.

4. The Scope and Limitations of Newton's Law

The collection of hypotheses on the laws of mechanics stated in section 1 forms together with Newton's law a coherent set of axioms whose consequences constitute Newtonian mechanics. These are studied in celestial mechanics. Even apart from any application to real physical problems, it is conceivable that a branch of mathematics should be devoted to these studies. However, it is interesting to see in what measure these axioms describe physical reality. It is in fact just because observations have for so long confirmed all the consequences of these hypotheses that so much work has been published in celestial mechanics. Phenomena as diverse as satellite and planetary motions, tides, precession of the Earth's axis, the motion of double stars, lunar libration, the gravity of the Earth, the motion of comets, etc., are all subject to a theory in celestial mechanics. Within the limits of the accuracy of measurements and calculations, we have (with one exception) always been able to verify the physical validity of the above hypotheses. We are thus led to consider Newton's law as the law of universal gravitation. The exception mentioned above was first noticed by Le Verrier in the motion of the planet Mercury. The precession of the perihelion (the point of Mercury's closest approach to the sun) is advanced by about 42″ per century over the value predicted by Newtonian mechanics. Similar discrepancies have since been observed for Venus, Earth, and Mars. However, for these last three planets the differences are of the same order of magnitude as observational errors, and even for Mercury the additional advance is much smaller than the advance already predicted by classical theories. Newton's law thus leads to a very good approximation, though a slight correction is still necessary.

Einstein was led to modify considerably the axiomatics of space and time. He did so for reasons which are largely outside the scope of celestial mechanics proper. Even a summary treatment of the theory of general relativity would exceed the aim of this book; basically, it denies the existence of absolute time and of a (Galilean) inertial system, postulates transmission of forces at the velocity of light, and considers that attractive forces are due to the fact that matter deforms the space-time continuum by its presence. However, in spite of such fundamental differences, relativity reduces in a first approximation to Newtonian mechanics. In a second approximation, the differences are negligible in view of the accuracy of measurements, the exception being the precessions of perihelions mentioned above.

The practical consequence of this situation is that we can continue to use the axioms of Newtonian mechanics; they are much easier to use than those of general relativity and may be corrected to second order for general relativity. One can easily show that these corrections are small, and that secondly, because of the small difference between the simplified motion of two bodies and the actual motion, these relativistic corrections may simply be added to the motion described by Newtonian mechanics.

In what follows we shall accept the laws and hypotheses already stated, bearing in mind that in a small number of cases small relativistic corrections should be made. This will only be necessary in the theory of certain planets,

5. The N-Body Problem

As an illustration of the preceding points we shall consider the N-body problem; the latter covers those problems where we wish to find the trajectories in time of N point masses whose only interaction is Newtonian gravitation. The interest of this problem (which is far from being solved) is that motion of the various bodies in the solar system represents a special case with small N, despite the fact that the Sun, planets and satellites are quasi-spherical and not point masses. In Chapter V we shall show that the mutual attraction of two of these bodies at sufficiently large distances can be calculated with high precision by assuming that the mass of each planet is concentrated at its centre of gravity.

Under the same conditions, for a quasi-spherical planet P acted on by distant point masses, the theorem on the motion of the centre of gravity allows us to calculate the forces on the various particles of the planet as if the latter's mass were concentrated at its centre of gravity.

These hypotheses are verified by the bodies in the solar system, with the exception of a few satellites close to their primary; we may thus replace all these bodies by their centres of gravity, reducing the problem of their motion to an N-body problem.

6. Equations of the N-Body Problem

Consider N point masses P_i, where $1 \le i \le N$, of masses m_i and coefficients x_i, y_i and z_i in an inertial reference frame with origin O.

We suppose that the P_j-th particle is gravitationally attracted by the $N-1$ other particles $P_i(1 \le i \le N; i \ne j)$. The resultant force on P_j is:

$$\mathbf{F_j} = \sum_{\substack{i=1 \\ i \ne j}}^{N} \frac{km_j m_i \mathbf{P_j P_i}}{|P_j P_i|^3} \tag{5}$$

The fundamental law of dynamics becomes:

$$m_j \frac{d^2 \mathbf{OP_j}}{dt^2} = \mathbf{F_j}$$

where $d^2\mathbf{OP_j}/dt^2$ is the acceleration of P_j. Eliminating m_j, we have:

$$\frac{d^2 \mathbf{OP_j}}{dt^2} = \sum_{\substack{i=1 \\ i \ne j}}^{N} \frac{km_i \mathbf{P_j P_i}}{|P_j P_i|^3} \quad 1 \le j \le N \tag{6}$$

We thus have N differential vector equations for the trajectory. If we project each of these vector equations along the three coordinate axes, we obtain $3N$ second-order differential equations which form a system of order $6N$:

$$\frac{d^2 x_j}{dt^2} = \sum_{\substack{i=1 \\ i \ne j}}^{N} \frac{km_i(x_i - x_j)}{[(x_i - x_j)^2 + (y_i - y_j)^2 + (z_i - z_j)^2]^{3/2}}$$

$$\frac{d^2 y_j}{dt^2} = \sum_{\substack{i=1 \\ i \ne j}}^{N} \frac{km_i(y_i - y_j)}{[(x_i - x_j)^2 + (y_i - y_j)^2 + (z_i - z_j)^2]^{3/2}} \tag{7}$$

$$\frac{d^2 z_j}{dt^2} = \sum_{\substack{i=1 \\ i \ne j}}^{N} \frac{km_i(z_i - z_j)}{[(x_i - x_j)^2 + (y_i - y_j)^2 + (z_i - z_j)^2]^{3/2}}$$

where $j = 1, 2, ..., N$.

Solution of this system of equations constitutes one of the most important branches of celestial mechanics: the dynamics of the solar system where each point P is identified with a body in the system.

7. Integrals of the N-Body Problem

An integral of a system of differential equations is a relation between the coordinates of the system, certain of their derivatives, and possibly time; it is valid for any time and depends on an arbitrary parameter. If one integral is known, the order of the system is reduced by one unit. We shall try to find the integrals of system (7).

(a) We apply the theorem on the motion of the centre of gravity to the whole N-body system. As it is subject to no *external* force, the motion of its centre of gravity is uniform and rectilinear with respect to an inertial frame. The coordinates of the centre of gravity are given by the vector relation

$$\mathbf{OG} = \sum_{j=1}^{N} \frac{m_j \mathbf{OP}_j}{\Sigma m_j}$$

Let $M = \Sigma m_j$ be the mass of the system.

We express the fact that the acceleration of G is zero:

$$\frac{d^2 \mathbf{OG}}{dt^2} = \frac{1}{M} \sum_{j=1}^{N} m_j \frac{d^2 \mathbf{OP}_j}{dt^2} = 0$$

Integrating the summation twice with respect to t, we obtain

$$\sum_{j=1}^{N} m_j \mathbf{OP}_j = \mathbf{A}t + \mathbf{B}$$

where $\mathbf{A}$ and $\mathbf{B}$ are two arbitrary vectors. Taking the coordinate components of this equation, we obtain the six integrals (three equations, each depending on two arbitrary parameters a_x, b_x or a_y, b_y or a_z, b_z):

$$\left. \begin{aligned} \sum_{j=1}^{N} m_j x_j &= a_x t + b_x \\ \sum_{j=1}^{N} m_j y_j &= a_y t + b_y \\ \sum_{j=1}^{N} m_j z_j &= a_z t + b_z \end{aligned} \right\} \tag{8}$$

(b) We apply the theorem on angular momentum to the N-body system. As there is no external force on the system, the moment of these forces about O is zero; the time derivative of the angular momentum of the system about O is consequently also zero. The angular momentum is thus constant. We have:

$$\sum_{j=1}^{N} \mathbf{OP_j} \wedge m_j \frac{d\mathbf{OP_j}}{dt} = \mathbf{C}$$

where $\mathbf{C}$ is an arbitrary constant vector with components C_x, C_y, C_z.
Taking the components of this equation, we have:

$$\left.\begin{array}{l} \displaystyle\sum_{j=1}^{N} m_j\left(y_j \frac{dz_j}{dt} - z_j \frac{dy_j}{dt}\right) = C_x \\[2mm] \displaystyle\sum_{j=1}^{N} m_j\left(z_j \frac{dx_j}{dt} - x_j \frac{dz_j}{dt}\right) = C_y \\[2mm] \displaystyle\sum_{j=1}^{N} m_j\left(x_j \frac{dy_j}{dt} - y_j \frac{dx_j}{dt}\right) = C_z \end{array}\right\} \tag{9}$$

We thus have three new integrals.

(c) We note that if we put

$$U = k \sum_{\substack{j=1 \\ i \neq j}}^{N} \sum_{i=1}^{N} \frac{m_i m_j}{[(x_i - x_j)^2 + (y_i - y_j)^2 + (z_i - z_j)^2]^{1/2}} \tag{10}$$

then, for a given h

$$\frac{\partial U}{\partial x_h} = k \sum_{\substack{i=1 \\ i \neq h}}^{N} \frac{m_i m_h (x_i - x_h)}{[(x_h - x_i)^2 + (y_h - y_i)^2 + (z_h - z_i)^2]^{3/2}}$$

The right hand side is identical with those of (7) with h in place of j.
We can thus also write system (7) as

$$m_j \frac{d^2 x_j}{dt^2} = \frac{\partial U}{\partial x_j}; \qquad m_j \frac{d^2 y_j}{dt^2} = \frac{\partial U}{\partial y_j}; \qquad m_j \frac{d^2 z_j}{dt^2} = \frac{\partial U}{\partial z_j} \tag{11}$$

Since U does not depend explicitly on time.

$$\frac{dU}{dt} = \sum_{j=1}^{N} \left(\frac{\partial U}{\partial x_j} \frac{dx_j}{dt} + \frac{\partial U}{\partial y_j} \frac{dy_j}{dt} + \frac{\partial U}{\partial z_j} \frac{dz_j}{dt}\right)$$

Using (11), this equation becomes

$$\frac{dU}{dt} = \sum_{j=1}^{N} \left(m_j \frac{d^2 x_j}{dt^2} \frac{dx_j}{dt} + m_j \frac{d^2 y_j}{dt^2} \frac{dy_j}{dt} + m_j \frac{d^2 z_j}{dt^2} \frac{dz_j}{dt}\right)$$

$$= \frac{1}{2} \sum_{j=1}^{N} m_j \frac{d}{dt}\left[\left(\frac{dx_j}{dt}\right)^2 + \left(\frac{dy_j}{dt}\right)^2 + \left(\frac{dz_j}{dt}\right)^2\right]$$

Integration gives an integral which, in fact, could have been deduced from the kinetic energy theorem,

$$\frac{1}{2} \sum_{j=1}^{N} m_j \left[\left(\frac{dx_j}{dt} \right)^2 + \left(\frac{dy_j}{dt} \right)^2 + \left(\frac{dz_j}{dt} \right)^2 \right] = U + h$$

where h is an arbitrary constant.

We have thus found ten integrals of the system (7). It is easily demonstrated that they are independent. Poincaré has shown that there are no other uniform ones.

Using these classical results on integrals, we see that they reduce the order of system (7) to $6N-10$. In particular, the famous two-body problem is reduced to a second order system which is completely integrable, as we shall see in the next Chapter.

CHAPTER II

THE TWO-BODY PROBLEM

8. The Importance of the Two-Body Problem

When we compare the various forces acting on a planet, we find that, at equal distances, the Sun to which we ascribe a mass M, has an attractive force M/m times as great as a planet of mass m. Table I gives the masses, distances from the Sun and numbers of satellites for the principal planets.

TABLE I

Name	Mean distance from Sun (Earth = 1)	Mass in millionths of the Sun's mass	Number of satellites
Sun		10^6	
Mercury	0.387	0.17	0
Venus	0.723	2.45	0
Earth	1.000	3.00	1*
Mars	1.524	0.32	2
Asteroids**	2 to 5.2	negligible	
Jupiter	5.203	954.8	12
Saturn	9.555	285.6	9 and rings
Uranus	19.218	43.7	5
Neptune	30.110	51.8	2
Pluto	39.600	2.7	0

* Plus an ever-increasing number of artificial satellites.
** 1650 such objects are now known, and new ones are discovered each year.

Cursory inspection of this table shows that the mean force exerted by the sun must be predominant for all the planets. The sun is at least 1000 times more massive than the largest planet, and the difference is very much greater for the smaller planets. If we neglect the attractive forces of the other planets as a first approximation, the motion of each planet can be studied by considering only the Sun and the planet in question: this is therefore a two-body problem. The variation from such motion, caused by the influence of other planets, is the object of planetary theory and is treated in Chapter VII.

The same approximation is valid for the motion of a satellite around its primary, though in this case the perturbations are more important since the relatively very much greater distance of the Sun does not always compensate for its large mass (Chapter VI).

A solution of the two-body problem often represents physical reality in an acceptable way, but this is not the main reason for the importance of this problem. Thus we shall see that all the most complete theories of celestial motion use functions appearing in solutions of two-body problems (elliptic case) as elementary functions. Solution of the two-body problem constitutes the basic algebra of the dynamics of the solar system – hence its importance in celestial mechanics.

9. Absolute and Relative Motion of Two Bodies

The equations of motion for two bodies are obtained by keeping only the coordinates of P_1 and P_2 of equations (6) from the preceding Chapter. The centre of gravity G may be taken as the origin of an inertial frame, since the theorem on the motion of G tells us that it is not accelerated. In this case, the equations are:

$$\frac{d^2 \mathbf{GP}_1}{dt^2} = \frac{km_2 \mathbf{P}_1 \mathbf{P}_2}{|P_2 P_1|^3}; \quad \frac{d^2 \mathbf{GP}_2}{dt^2} = \frac{km_1 \mathbf{P}_2 \mathbf{P}_1}{|P_2 P_1|^3} \tag{1}$$

From the definition of G, we have:

$$m_1 \mathbf{GP}_1 + m_2 \mathbf{GP}_2 = 0$$

from which we deduce

$$\mathbf{P}_1 \mathbf{P}_2 = -\frac{m_1 + m_2}{m_2} \mathbf{GP}_1 = \frac{m_1 + m_2}{m_1} \mathbf{GP}_2 \tag{2}$$

These expressions, substituted in (1), give:

$$\frac{d^2 \mathbf{GP}_1}{dt^2} = \frac{-km_2^3}{(m_1 + m_2)^2} \frac{\mathbf{GP}_1}{|GP_1|^3}; \quad \frac{d^2 \mathbf{GP}_2}{dt^2} = \frac{-km_1^3}{(m_1 + m_2)^2} \frac{\mathbf{GP}_2}{|GP_2|^3}.$$

RESULT: *Each body is attracted towards the centre of gravity as if a mass $M = [m'^3/(m+m')^2]$ were concentrated there, m and m' being the masses of the two bodies.*

In practice, the centre of gravity of two bodies is not accessible to measurement, which has to be made with respect to some material point. Let P_1 be this point, and let the motion of P_2 be referred to a system of axes with P_1 as the origin, these axes always remaining parallel to the inertial frame.

Eliminating $\mathbf{GP}_1$ or $\mathbf{GP}_2$ from (1) and (2), we obtain

$$\frac{d^2 \mathbf{P}_1 \mathbf{P}_2}{dt^2} = \frac{-k(m_1 + m_2) \mathbf{P}_1 \mathbf{P}_2}{|P_1 P_2|^3} \tag{3}$$

In the two-body problem, the motion of one body relative to the other, referred to axes parallel to an inertial frame, is that which would be due to the attraction exerted by a central mass, equal to the sum of the masses of the two bodies, acting on the other mass.

As could be seen from equation (2), the absolute and relative motions are homothetic. Only the relative motion will be treated further in subsequent discussion.

10. Form of the Trajectories

Consider a particle P of mass m attracted by a particle A of mass M according to Newton's law. From the preceding discussion, the force in a coordinate system with A as origin and parallel to an inertial frame is

$$F = \frac{-k(M+m)\,m}{r^2}$$

We put $\mu = k(M+m)$. The acceleration of P is then $-\mu/r^2$.

Let us apply the fundamental theorems to this motion.

(a) The theorem on angular momentum becomes $\mathbf{AP} \wedge \mathbf{V_P} = \boldsymbol{\sigma}$, a constant vector (Figure 1). Let r and θ be the polar coordinates of P; we know from Section 2 that the motion is in a plane. The radial and tangential components of the velocity of P, $\mathbf{V_P}$, are then dr/dt and $r\,d\theta/dt$.

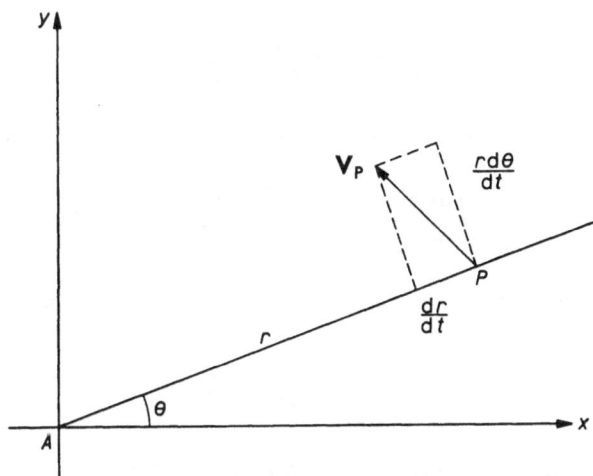

Fig. 1.

We call the constant magnitude of $\boldsymbol{\sigma}$, the *constant of areas* C:

$$r \times \frac{r\,d\theta}{dt} = r^2\frac{d\theta}{dt} = C \tag{4}$$

(b) The kinetic energy theorem in differential form is, with V as the magnitude of $\mathbf{V_P}$:

$$d(\tfrac{1}{2}mV^2) - \mathbf{F}\cdot\mathbf{V_P}\,dt = 0,$$

or

$$\tfrac{1}{2}mV^2 - \int \frac{-\mu m}{r^2}\frac{dr}{dt} \times dt = C'$$

where C' is a constant and the scalar product $\mathbf{F} \cdot \mathbf{V_P}$ has been replaced by the product of the magnitude of F and the projection of V_P along the radius vector. We thus obtain:

$$\tfrac{1}{2}mV^2 - \frac{\mu m}{r} = C'$$

Finally, putting $h = C'/m$, the energy constant, we have

$$V^2 = 2h + \frac{2\mu}{r} \tag{5}$$

(c) Elimination of the time t from equations (4) and (5) yields

$$V^2 = \frac{dr^2 + r^2\,d\theta^2}{dt^2} = \frac{C^2(dr^2 + r^2\,d\theta^2)}{r^4\,d\theta^2},$$

which on substitution in (5) gives the differential equation of the trajectory:

$$\frac{dr^2}{r^4\,d\theta^2} + \frac{1}{r^2} - \frac{2\mu}{C^2 r} - \frac{2h}{C^2} = 0 \tag{6}$$

Putting $u = 1/r - \mu/C^2$ we obtain $du = -dr/r^2$. In terms of this new variable, eq. (6) becomes

$$\left(\frac{du}{d\theta}\right)^2 + u^2 - \left(\frac{\mu^2}{C^4} + \frac{2h}{C^2}\right) = 0$$

We call the quantity in the bracket H^2, and suppose that it is positive (if not, there will be no real trajectory). Then

$$\left(\frac{du}{d\theta}\right)^2 = H^2 - u^2$$

which is immediately integrable, giving $u = H \cos(\theta - \theta_0)$, where θ_0 is an arbitrary constant. Alternatively, in terms of the radius vector r,

$$\frac{1}{r} = \frac{\mu}{C^2}\left[1 + \sqrt{1 + \frac{2C^2 h}{\mu^2}}\,\cos(\theta - \theta_0)\right] \tag{7}$$

which is the equation of a conic section with focus A.

(d) Let us put

$$\left.\begin{array}{l} p = \dfrac{C^2}{\mu} \\[4mm] e = \sqrt{1 + \dfrac{2C^2 h}{\mu^2}} = \sqrt{1 + \dfrac{2hp}{\mu}} \\[4mm] v = \theta - \theta_0 \end{array}\right\} \tag{8}$$

whence

$$2h = \frac{\mu(e^2 - 1)}{p}$$

The equation of the conic section then becomes

$$\frac{1}{r} = \frac{1 + e \cos v}{p}$$

The eccentricity of the conic is e and its parameter p. The semi-major axis a is given by $p = a(1 - e^2)$. The angle v, called the *true anomaly*, is measured from the point of nearest approach to A. This point is called the *perifocus**. The distance to the perifocus is obtained by setting $v = 0$ and is thus $a(1 - e)$. The other orbital point on the major axis is the apofocus (or apogee, or aphelion); the distance to the apofocus is $a(1 + e)$.

(e) We can discuss the nature of this conic section. From (8), it is an ellipse, a parabola, or a branch of the hyperbola, according to whether h is negative, zero, or positive. If V_0 is the velocity at the original instant for which the radius vector is r_0, in accordance with (5)

$$2h = V_0^2 - \frac{2\mu}{r_0},$$

we deduce the following results:

1) If $V_0 = \sqrt{2\mu/r_0} = V_p$ (called the parabolic velocity): a parabolic orbit;
2) If $V_0 > V_p$: a hyperbolic orbit;
3) If $V_0 < V_p$: an elliptical orbit.

11. Kepler's Laws

Kepler's laws state:

1) The planets move in plane curves, and their radius vectors sweep out equal areas in equal times.
2) The orbits of the planets are ellipses with the Sun at one of the foci.
3) The squares of the periods of revolution of planets around the Sun are to each other as the cubes of the semi-major axes of their orbits.

The first two laws have already been proved, on the assumption that we may consider each planet independently in its motion around the Sun. Moreover, the first law is a consequence of the angular momentum theorem in an arbitrary central field of force. The energy constant h is negative for the planets.

The third law is proved as follows:

---

* This is known as the "perigee" if the central body is the Earth, and "perihelion" if the central body is the Sun.

The area of an ellipse is $\pi a^2\sqrt{1-e^2}$. Let P be the period of revolution. The constant of areas C is such that

$$\frac{1}{2}\int_0^P C\,dt = \frac{CP}{2} = \pi a^2 \sqrt{1-e^2}$$

$$C = \frac{2\pi}{P}a^2 \sqrt{1-e^2}. \tag{9}$$

We have already seen that $C^2 = p\mu = \mu a(1-e^2)$.
Eliminating C from these last two equations, we find

$$\frac{4\pi^2}{P^2}a^2 = \mu \tag{10}$$

If we neglect the planetary mass, $\mu = kM$ is the same for every planet, and the third law is proved. We note that with the actual masses given in Table I, this law holds only to within 0.1 % in the case of Jupiter.

12. Study of Elliptical Motion

(a) We introduce a new angle (Figure 2) to study the elliptical motion ($h<0$) of a particle as a function of time. Let O, F, and P be respectively the centre, the focus, and perifocus of the ellipse. M is a point on the ellipse whose coordinates on the axes Fx and Fy are $r\cos v$ and $r\sin v$. Let M' be that point on the auxiliary circle $(OM'=a)$ which projects the same point H as M on Ox, and is such that M and M' are on the same side of Ox. The angle $E(\mathbf{OP}, \mathbf{OM'})$* is called the *eccentric anomaly*.

We know that the ellipse is an affine transform of the auxiliary circle in the ratio $\sqrt{1-e^2}$, i.e. we always have $\overline{HM}=\sqrt{1-e^2}\,HM'$. Since $\overline{OF}=ae$, we can calculate the

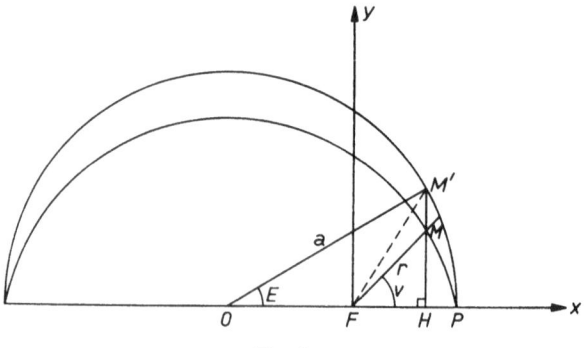

Fig. 2.

---

* Often denoted by u. E is the internationally recommended symbol.

projection of *FM* on *Fx* and *Fy* in two different ways:

$$\left.\begin{array}{l} x = r \cos v = a(\cos E - e) \\ y = r \sin v = a\sqrt{1 - e^2}\, \sin E \end{array}\right\} \tag{11}$$

From this, by calculating r^2:

$$r = a(1 - e \cos E) \tag{12}$$

Eliminating *r*, we have

$$\cos v = \frac{\cos E - e}{1 - e \cos E}$$

and calculating $\tan^2(v/2) = (1 - \cos v)/(1 + \cos v)$, we find

$$\tan^2\frac{v}{2} = \frac{1 - e \cos E - \cos E + e}{1 - e \cos E + \cos E - e} = \frac{1 + e}{1 - e}\frac{1 - \cos E}{1 + \cos E} = \frac{1 + e}{1 - e}\tan^2\frac{E}{2}$$

Since *v* and *E* are in the same half-plane, *v*/2 and *E*/2 are in the same quadrant and we finally have:

$$\tan\frac{v}{2} = \sqrt{\frac{1 + e}{1 - e}}\,\tan\frac{E}{2} \tag{13}$$

which relates the true anomaly to the eccentric anomaly.

(b) A third "anomaly"* appears in our treatment:

The *mean anomaly*, defined as $M = n(t - t_0)$, where the mean motion $n = 2\pi/P$ (the period *P* is as defined previously), *t* is the time of computation, and t_0 is the time when the body passes through the perifocus.

(c) With the notation *n*, we note that, from (8) and $p = a(1 - e^2)$ we have $h = -\mu/2a$. The standard formulae which we have derived then become:

$$C = r^2\frac{dv}{dt} = na^2\sqrt{1 - e^2}: \tag{14}$$

angular-momentum integral (9);

$$V^2 = \mu\left(\frac{2}{r} - \frac{1}{a}\right): \tag{15}$$

energy integral (5);

$$n^2a^3 = \mu: \tag{16}$$

Kepler's third law (10).

(d) We are now in a position to study changes in *E* as a function of *t*.

---

* The word "anomaly" as used in celestial mechanics and astronomy means that angle which is zero when the body passes through the perifocus; the word "longitude" means an angle referred to an arbitrary fixed origin.

The area swept out in time $t - t_0$ is the area PMF (Figure 2).

$$\text{area } (PMF) = \tfrac{1}{2} \int_{t_0}^{t} C \, dt = \tfrac{1}{2} na^2 \sqrt{1 - e^2} \, (t - t_0) = \tfrac{1}{2} a^2 \sqrt{1 - e^2} \, M$$

M, and later E, are assumed to be in radians.

We deduce area $PM'F$ from the preceding one by using the affine relationship between the ellipse and the circle:

$$\text{Area } (PM'F) = \frac{1}{\sqrt{1 - e^2}} \; \text{Area } (PMF) = \tfrac{1}{2} a^2 M$$

However,

$$\text{Area } (PM'F) = \text{Area } (PM'O) - \text{Area } (FM'O) \tag{17}$$

$PM'O$ is a circular sector of area $\tfrac{1}{2} a^2 E$.

$FM'O$ is a triangle of base $FO = ae$ and height $HM' = a \sin E$. Its area is thus $\tfrac{1}{2} a^2 e \sin E$. Substituting these results in (17), we have

$$\tfrac{1}{2} a^2 M = \tfrac{1}{2} a^2 E - \tfrac{1}{2} a^2 e \sin E$$

or

$$E - e \sin E = M = n(t - t_0) \tag{18}$$

which is Kepler's equation between E and t.

This equation may also be obtained by differentiating (13), substituting the result in (14), eliminating v and r by means of (11) and (12), and integrating the result.

Equations (13) and (18) relate the three anomalies to one another, and allow us to calculate each at any time t. Consequently, we can find the coordinates of the body by using (11); this completely solves the problem of elliptical motion.

13. The Orbital Elements

Knowledge of the quantities $r \sin v$ and $r \cos v$ is not generally sufficient to specify the spatial position of the body with respect to a set of arbitrary axes. We thus define certain conventional parameters to fix the orbit in a set of cartesian axes.

We shall assume that the origin of these axes is the central body (for relative motion) or the centre of gravity of the two bodies (for absolute motion). Let us call the XOY plane the principal plane (Figure 3). Its normal, OZ, is the direction of the pole of the principal plane. In practice, the principal plane is either the terrestrial equator or the plane of the ecliptic.

The intersection of the orbital plane with the principal plane is the *line of nodes*. This line intersects the orbit in two points. At the *ascending node* (N), the z-coordinate of the body increases. The other intersection constitutes the descending node. The direction ON of the ascending node is defined by $\Omega = (OX, ON)$, to $360°$; Ω is called the *longitude of the ascending node*.

The angle i between the principal plane and the orbital plane is the *inclination*, and varies between 0 and $180°$. If the sense of the revolution projected onto the principal plane is direct, then $0° < i < 90°$; if it is retrograde, $90° < i < 180°$.

Ω and i thus define the orbital plane and the sense of revolution of M is this plane.

The orbit in this plane is a conic section with focus O. Its shape is defined by its *eccentricity e*, and its size by its *semi-major axis a* (or by the parameter $p = a(1 - e^2)$ in the case of parabolic orbits). For a hyperbola, $a < 0$.

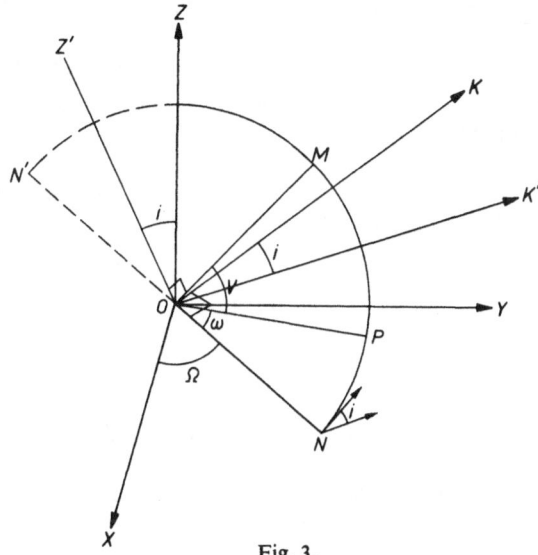

Fig. 3.

We still have to fix the position of the conic section in this plane. This can be done by specifying the direction of the perifocus, OP. The angle $\omega - (ON, OP)$ is called the *argument of the perifocus*; it is measured along the direction of motion and defines uniquely the position of the orbit in its plane.

Finally, we have seen that to specify the position of M in its orbit we use the true anomaly v. To calculate it as a function of time, we solve Kepler's equation after fixing a sixth parameter: t_0, the *time of perifocus passage* or *the epoch*.

The six parameters a, e, i, Ω, ω, and t_0 uniquely specify the motion of M in its orbit as a function of time.

NOTES:

1. The *longitude of the perigee* ϖ is often used instead of the *argument of the perigee* ω. It is defined as $\varpi = \omega + \Omega$ and is the sum of two angles not in the same plane. However, if $i \to 0$, ON becomes indeterminate and hence ω and Ω, while ϖ tend to a definite limit.

2. The mean motion n is not a seventh orbital element. It is a function of a: $n^2 a^3 = \mu$.

14. Cartesian Coordinates of the Body

Let us compute the projections of OM on the axes OX, OY, and OZ. The coordinates of M are $r\cos(\omega+v)$ and $r\sin(\omega+v)$ with respect to axes ON and OK in the orbital plane (Figure 3).

We rotate the axes by $-i$ about ON; the coordinates with respect to ON, OK', and OZ are

$$r\cos(\omega + v), \qquad r\sin(\omega + v)\cos i, \qquad r\sin(\omega + v)\sin i$$

Finally, rotation of $-\Omega$ about OZ provides the required coordinates

$$\left.\begin{aligned}
X &= r[\cos(\omega + v)\cos\Omega - \sin(\omega + v)\sin\Omega\cos i]\\
Y &= r[\cos(\omega + v)\sin\Omega + \sin(\omega + v)\cos\Omega\cos i]\\
Z &= r\sin(\omega + v)\sin i
\end{aligned}\right\} \qquad (19)$$

Table II below gives $r\sin v$ and $r\cos v$ (the hyperbolic and parabolic cases are given only for reference and are not treated here).

TABLE II

Quantity	Elliptic case	Hyperbolic case	Parabolic case
$r\cos v$	$a(\cos E - e)$	$a(e - \operatorname{ch} F)$	$\dfrac{p}{2}(1 - s^2)$
$r\sin v$	$a\sqrt{1 - e^2}\sin E$	$a\sqrt{e^2 - 1}\operatorname{sh} F$	ps
$t - t_0$	$\dfrac{1}{n}(E - e\sin E)$	$\dfrac{1}{n}(-F + e\operatorname{sh} F)$	$\dfrac{p^{3/2}}{\sqrt{\mu}}\left(\dfrac{s}{2} + \dfrac{s^3}{6}\right)$

We can also ask the inverse question: having observed various positions of a body, how can we determine the elements of its orbit?

If we know its Cartesian coordinates, two positions will suffice to solve the preceding equations. In general, however, astronomical observations give not distances but directions. In this case, at least three observations are needed. Many methods have been used to solve this problem, but their description is not within the scope of this book. The principles of such methods are described by many authors.*

15. Astronomical Units in the Solar System

The accuracy with which the semi-major axis of an orbit can be defined depends essentially on the value of the period and on our knowledge of the quantity μ; we

---

* See WATSON, *Theoretical Astronomy*, Dover 1964, and also Andoyer, and Danjon.

have, from (10):

$$\frac{4\pi^2}{P^2} a^3 = \mu = k(M + m)$$

Planetary observations give P with great accuracy (relative error $10^{-8}-10^{-9}$). If we take the C.G.S. value of k as given in Section 3, only to three decimal places, we shall lose the high accuracy of the observations.

All calculations on the solar system are therefore done in units independent of the C.G.S. system. The unit of time remains the same, and the masses of all the planets are expressed in terms of the mass of the Sun, taken as unity ($M=1$). Such masses are listed in Table I. Finally, it is preferred to define the unit of length indirectly, by fixing an assuming value for k:

$$\sqrt{k} = 0.01720,20989,5000...; \quad k = 0.00029,59122,08266...$$

Thus, for $a=1$, a planet of negligible mass has a period $P=2\pi/\sqrt{k}$.

The astronomical unit of length is the semi-major axis of the orbit of an unperturbed planet with a negligible mass, whose period around the Sun is 365,2568983263 days.

The semi-major axis of the Earth's orbit differs from the astronomical unit by only a few parts in 10^{-7}. This unit would thus be unaffected by any improvement in the theory of the Earth's motion.

CHAPTER III

SYSTEMS OF CANONICAL EQUATIONS

Having solved the two-body problem in the last chapter, we shall now examine some properties of the differential equations that occur in the three-body problem. However, before approaching this task we shall cast these equations into their most convenient form.

16. N-Body Problem Equations in a Relative Frame of Reference

As in the case of the two-body problem and for the same reason, we study the motion of an N-body system (such as the solar system) in a frame which is centred on one of the bodies P_1 (for example the Sun) and which is parallel to an inertial frame. The order of the system is thus reduced by six. This is equivalent to using the information provided by the centre of gravity theorem, which no longer applies to the relative system.

Using the notation of Chapter I, the coordinates of a body P_j are:

$$X_j = x_j - x_1; \quad Y_j = y_j - y_1; \quad Z_j = z_j - z_1$$

Also, let

$$\Delta_{ij} = |P_i P_j|$$

The equations of motion of P_j with respect to P_1 are obtained by subtracting term by term the equations (7) of Chapter I for these two bodies:

$$\frac{d^2 X_j}{dt^2} = -k(m_1 + m_j)\frac{X_j}{\Delta_{j1}^3} + \sum_{\substack{i=2 \\ i \neq j}}^{N} km_i \left(\frac{X_i - X_j}{\Delta_{ij}^3} - \frac{X_i}{\Delta_{1i}^3} \right) \quad ij = 2, 3, ..., N \quad (1)$$

with two similar equations in a Y and Z.

By calculating the partial derivatives of the functions

$$V_j = \frac{k(m_1 + m_j)}{\Delta_{j1}} + \sum_{\substack{i=2 \\ i \neq j}}^{N} km_i \left(\frac{1}{\Delta_{ij}} - \frac{X_i X_j + Y_i Y_j + Z_i Z_j}{\Delta_{1i}^3} \right) \quad (2)$$

we see that the system given by (1) becomes simply

$$\frac{d^2 X_j}{dt^2} = \frac{\partial V_j}{\partial X_j}; \quad \frac{d^2 Y_j}{dt^2} = \frac{\partial V_j}{\partial Y_j}; \quad \frac{d^2 Z_j}{dt^2} = \frac{\partial V_j}{\partial Z_j} \quad (3)$$

NOTE:

Notice the difference from the results of section 7: there are as many functions V_j as there are bodies minus one, but each is simpler than the function U.

17. Reductions of the Equations for a Three-Body Problem

There is a set of axes which combines the advantages of the two preceding systems: the coordinates of each body are referred to a point independent of the position of that body, and there is a unique function V. We shall define these axes only for the case of three-bodies, but the results can be extended to N bodies.

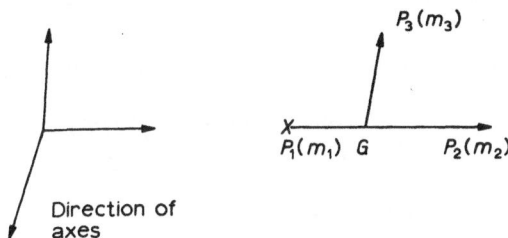

Fig. 4.

We call the coordinates of P_2 with respect to P_1: x, y, z $(x=X_2$, etc.$)$; x', y', z' are the coordinates of P_3 with respect to the centre of gravity G of P_1 and P_2. The two sets of axes stay parallel (Figure 4). Then

$$x' = X_3 - \frac{m_2}{m_1 + m_2} X_2; \quad y' = Y_3 - \frac{m_2}{m_1 + m_2} Y_2;$$

$$z' = Z_3 - \frac{m_2}{m_1 + m_2} Z_2$$

and

$$\Delta_{13}^2 = \sum \left(x' + \frac{m_2}{m_1 + m_2} x \right)^2;$$

$$\Delta_{23}^2 = \sum \left(x' - \frac{m_1}{m_1 + m_2} x \right)^2$$

The equations relative to P_2 are

$$\frac{d^2 x}{dt^2} = \frac{-k(m_1 + m_2) x}{\Delta_{12}^3} + km_3 \left[\frac{x' - \frac{m_1}{m_1 + m_2} x}{\Delta_{23}^3} - \frac{x' + \frac{m_2}{m_1 + m_2} x}{\Delta_{13}^3} \right] \qquad (4)$$

and two analogous equations for y and z.

We can write similar equations in x', y', and z', starting from those in X_3, Y_3, and Z_3.

$$\frac{d^2x'}{dt^2} + \frac{m_2}{m_1 + m_2}\frac{d^2x}{dt^2} = -\frac{k(m_1 + m_3)\left(x' + \dfrac{m_2}{m_1 + m_2}x\right)}{\Delta_{13}^3}$$

$$- km_2\left[\frac{x' - \dfrac{m_1}{m_1 + m_2}x}{\Delta_{23}^3} + \frac{x}{\Delta_{12}^3}\right] \qquad (5)$$

and two analogous equations in y' and z'.

By subtracting these two equations term by term, we can eliminate d^2x/dt^2 from the second.

It can be shown that if, as in the analogous previous cases, we put

$$V = k\frac{m_1 m_2}{\Delta_{12}} + k\frac{m_1 m_3}{\Delta_{13}} + k\frac{m_2 m_3}{\Delta_{23}} \qquad (6)$$

the equations of the three-body problem become

$$\frac{m_1 m_2}{m_1 + m_2}\frac{d^2x}{dt^2} = \frac{\partial V}{\partial x}; \quad \frac{(m_1 + m_2)m_3}{m_1 + m_2 + m_3}\frac{d^2x'}{dt^2} = \frac{\partial V}{\partial x'} \qquad (7)$$

and four analogous equations in y and z.

18. The Case in which One of the Bodies has Negligible Mass

The mass of the body studied can often be neglected. For example, this is the situation in the case of the motion of an asteroid perturbed by Jupiter, or that of a satellite perturbed by the Sun. The motion of the perturbing body P_1 is not affected by the small body P_2. The positions of G and P_2 then coincide. The motion of P_3 is simply due to the action of P_1 – motion according to Kepler's laws.

It is therefore sufficient to consider only the equations in X_2. Under these conditions, there is no point in carrying out the preceding reduction. It is sufficient to consider equations (3) with $j=2$ and $i=1$, which become:

$$\frac{d^2X_2}{dt^2} = \frac{\partial V_2}{\partial X_2}, \quad \text{etc. ...;}$$

$$V_2 = \frac{k(m_1 + m_2)}{\Delta_{12}} + km_3\left(\frac{1}{\Delta_{23}} - \frac{X_2X_3 + Y_2Y_3 + Z_2Z_3}{\Delta_{13}^3}\right) \qquad (8)$$

19. Canonical Form of the Equations

The systems of equations given by (7) and (8), as well as (11) of Chapter I, have been brought into the following form:

$$m_j \frac{d^2 x_j}{dt^2} = \frac{\partial V}{\partial x_j} \qquad j = 1, 2, ..., n \tag{9}$$

Introducing n new variables $y_j = m_j (dx_j/dt)$ and putting

$$T = \frac{1}{2} \sum_{j=1}^{n} \frac{y_j^2}{m_j},$$

we obtain

$$\frac{\partial T}{\partial x_j} = 0; \qquad \frac{\partial T}{\partial y_j} = \frac{y_j}{m_j}; \qquad \frac{\partial V}{\partial y_j} = 0$$

We can write (9) as a system of linear equations of order $2n$.

$$\frac{dx_j}{dt} = \frac{\partial F}{\partial y_j}; \qquad \frac{dy_j}{dt} = -\frac{\partial F}{\partial x_j} \qquad j = 1, 2, ..., n \tag{10}$$

with $F = T - V$.

A system of equations of this form is called *canonical*. The function F, common to all the equations, is called the *Hamiltonian* or *characteristic function*, and x_j and y_j are *conjugate variables*. Such equations are very important and are encountered in many problems of celestial mechanics, quantum mechanics, and so on.

The function F may or may not be an explicit function of time, according to whether the positions of certain perturbing bodies are previously known or not. In particular, for the three-body case where one mass is negligible, the previous section shows that the canonical system is of sixth order; V is a function of t via the known positions of P_3.

20. The Case in which F is not a Function of t

The preceding result already shows the importance of canonical equations in celestial mechanics. We shall consider the most frequently used properties of such systems. However, it is convenient to show how one can eliminate the explicit presence of t in F.

Consider the system of $2n$ canonical equations:

$$\frac{dq_j}{dt} = \frac{\partial F}{\partial p_j}(q_i, p_i, t); \qquad \frac{dp_j}{dt} = -\frac{\partial F}{\partial q_j}(q_i, p_i, t); \qquad 1 \leqslant j, \; i \leqslant n \tag{11}$$

If F depends explicitly on t, $\partial F/\partial t \neq 0$.

We introduce two new variables q_{n+1} and p_{n+1}, where q_{n+1} replaces t as the explicit variable in F, t remaining the independent variable, and p_{n+1} is the variable conjugate to q_{n+1}. The Hamiltonian F does not contain p_{n+1}, but we can always add a function of p_{n+1} to it without affecting equations (11). We choose this function such that the new system

$$
\left.
\begin{aligned}
\frac{dq_j}{dt} &= \frac{\partial F^*(q_j, p_j, q_{n+1}, p_{n+1})}{\partial p_j}; \\
\frac{dp_j}{dt} &= -\frac{\partial F^*(q_i, p_j, q_{n+1}, p_{n+1})}{\partial q_j},
\end{aligned}
\right\}
\tag{12}
$$

with $j=1, 2, ..., n, n+1$, allows the solution $q_{n+1}=t$.

For this purpose, we must take $F^*=F+p_{n+1}$. The last two equations become

$$
\frac{dq_{n+1}}{dt} = \frac{\partial p_{n+1}}{\partial p_{n+1}} = 1;
$$

$$
\frac{dp_{n+1}}{dt} = \frac{-\partial F^*}{\partial q_{n+1}} = -\left(\frac{\partial F^*}{\partial t}\right)_{t=q_{n+1}}
$$

We have thus reduced the initial system to another canonical system, of order $2n+2$, whose characteristic function F^* no longer contains t. This new system is more general: it embraces the solution of the first system if we replace q_{n+1} by t and ignore p_{n+1}.

21. Integral of a System of Canonical Equations

Let us suppose, in the first place, that we have a system of $2n$ canonical equations whose characteristic function F does not depend on t.

$$
\frac{dq_j}{dt} = \frac{\partial F(q_i, p_i)}{\partial p_j}; \quad \frac{dp_j}{dt} = -\frac{\partial F(q_i, p_i)}{\partial q_j}; \quad 1 \le i, \ j \le n
\tag{13}
$$

Since F is not an explicit function of t, its total derivative is:

$$
\frac{dF}{dt} = \sum_{j=1}^{n} \left(\frac{\partial F}{\partial q_j} \frac{dq_j}{dt} + \frac{\partial F}{\partial p_j} \frac{dp_j}{dt} \right)
$$

If we substitute for the $2n$ functions q_j and p_j which constitute a solution of the system, then:

$$
\frac{dF}{dt} = \sum_{j=1}^{n} \left(\frac{\partial F}{\partial q_j} \frac{\partial F}{\partial p_j} - \frac{\partial F}{\partial p_j} \frac{\partial F}{\partial q_j} \right) = 0
$$

This equation can be integrated. The equality

$$
F(q_i, p_i) = C
$$

is satisfied for any solution $q_i(t), P_i(t)$ that we put in it, and it is thus an integral of equations (13).

In practice, the partial derivatives of F are also independent of time; we can eliminate t, which occurs only as the differential dt, and by choosing for example q_n as a new independent variable, write the system (13) in the form:

$$\frac{dq_j}{dq_n} = \frac{\partial F/\partial p_j}{\partial F/\partial p_n}; \quad 1 \leqslant j \leqslant n - 1;$$

$$\frac{dp_j}{dq_n} = -\frac{\partial F/\partial q_j}{\partial F/\partial p_n}; \quad 1 \leqslant j \leqslant n$$

This is a system of order $2n-1$, having $F = C$ as its integral. If we have one solution of the system, a function of q_n, then we can obtain t from the integration

$$t - t_0 = \int (dq_n/\partial F)/\partial p_n \tag{14}$$

since the right-hand side is a function of q_n alone.

The system is thus reduced to a canonical system of order $2n-2$, whose Hamiltonian depends on the independent variable and an integration. The methods generally used in celestial mechanics rarely call on this property; it is more often the inverse one, shown in the preceding section, which allows elimination of t from the characteristic function.

Application of this process to a system of order $2n+2$ whose characteristic function is F^* of section 20 gives $F^* = C$ as an integral and equation (14) reduces to $dt = dq_{n+1}$, since $\partial F^*/\partial p_{n+1} = 1$.

We return to the original system with no additional information, since the integral $F^* = C$ is a function of an extra variable p_{n+1}.

22. Canonical Transformations of Variables

Transformation of variables is one of the most frequently employed methods of solving the equations of celestial mechanics. This method, which we shall illustrate, proves particularly effective when the equations are written in canonical form. It consists essentially of transforming the variables p_i, $q_i (1 \leq i \leq N)$ into new variables P_i, $Q_i (1 \leq i \leq N)$, such that the equations written in these new variables are simpler. If, in addition the new system of equations is canonical, we say that the transformation of variables is *canonical*. If we succeed in finding such a transformation, we can continue the process until we have a system of equations which are easily integrated (see Chapter V, for example).

We shall now find the necessary and sufficient conditions under which a transformation of variables is canonical.

A. THE NECESSARY CONDITION

Consider the differential system (13):

$$\frac{dq_j}{dt} = \frac{\partial F(q_i, p_i)}{\partial p_j}; \quad \frac{dp_j}{dt} = -\frac{\partial F(q_i, p_i)}{\partial q_j} \quad 1 \leqslant i, \ j \leqslant N$$

where F does not depend explicitly on t, and let $P_j, Q_j, 1 \le j \le N$ be $2N$ new variables which are canonical too.

We examine the quantity

$$d\theta = \sum_j p_j \, dq_j - F \, dt \tag{15}$$

We have that $dp_j/dt = -\partial F/\partial q_j$ for all the values of j and that p_j depends only on t since it is a solution; we can also write

$$\frac{\partial p_j}{\partial t} = -\frac{\partial F}{\partial q_j} \tag{16}$$

We know that a necessary and sufficient condition for $\Sigma X_i \, dx_i$ to be a total differential is that the quantities $(\partial X_i/\partial x_k - \partial X_k/\partial x_i)$ are zero.

Applying this condition to the right-hand side of (15), and remembering that only F depends on the q_j and that the p_j depends on t alone, we obtain

$$\frac{\partial p_j}{\partial t} + \frac{\partial F}{\partial q_j} = 0,$$

i.e. system (16).

The quantity $\sum_j p_j \, dq_j - F dt$ is thus a total differential.

We can apply the same reasoning to the new canonical system

$$\frac{dQ_j}{dt} = \frac{\partial F^*(P_i, Q_i)}{\partial P_j}, \quad \frac{dP_j}{dt} = -\frac{\partial F^*(P_i, Q_i)}{\partial Q_j}$$

The new Hamiltonian F^* is not necessarily identical with F. Here again, $\sum P_j \, dQ_j - F^* \, dt = d\theta^*$, is a total differential. Subtracting term by term, we see that a necessary condition for the transformation of the variable to be canonical is

$$\sum_j P_j \, dQ_j - \sum_j p_j \, dq_j = d(\theta^* - \theta) + (F^* - F) \, dt$$

Putting $K = F^* - F$, this becomes

$$\sum_j P_j \, dQ_j - \sum_j p_j \, dq_j - K \, dt = dW \tag{17}$$

where K is a function of the variables and dW is a total differential.

B. THE SUFFICIENT CONDITION

We shall show that condition (17) is sufficient.

Since the initial system is canonical,

$$\sum_j p_j \, dq_j - F \, dt = d\theta$$

We substitute this in (17) and obtain

$$\sum_j P_j \, dQ_j - (F + K) \, dt = d(W + \theta) \tag{18}$$

The left-hand side is a total differential because the right-hand side is one. The condition for this is

$$\frac{\partial P_j}{\partial t} = -\frac{\partial (F + K)}{\partial Q_j}$$

Since we suppose that P_j is a variable, it depends only on t, this gives

$$\frac{dP_j}{dt} = -\frac{\partial (F + K)}{\partial Q_j} \tag{19}$$

But we also know that the total differential of $\sum_j P_j Q_j$ is

$$d\left(\sum_j P_j Q_j\right) = \Sigma P_j \, dQ_j + \Sigma Q_j \, dP_j.$$

Equation (18) can be written

$$d\left(\sum_j P_j Q_j\right) - \Sigma Q_j \, dP_j - (F + K) \, dt = d(W + \theta)$$

or

$$\Sigma Q_j \, dP_j + (F + K) \, dt = d\left(\sum_j P_j Q_j - W - \theta\right) \qquad .$$

Again, the left-hand side must be a total differential. As before, we see that this implies

$$\frac{dQ_j}{dt} = \frac{\partial (F + K)}{\partial P_j} \tag{20}$$

Equations (19) and (20) show that the system of equations in P_j and Q_j, having a Hamiltonian $F + K$, is canonical.

RESULT: We have shown that condition (17)

$$\sum_j P_j \, dQ_j - \Sigma p_j \, dq_j - K \, dt = dW$$

is a necessary and sufficient condition for the change of variables from p_j, q_j to P_j, Q_j to be canonical. The new characteristic function is $F + K$, regardless of the explicit presence of t in the Hamiltonians.

NOTE:

The transformation $(p_i, q_i) \rightarrow (P_i, Q_i)$ of (17) is called a contact transformation; it plays an important part in the theory of partial differential equations. Its classical properties will not, however, be of use here.

23. Examples of Canonical Transformations

A. CHANGE OF VARIABLES BY MEANS OF A DETERMINING FUNCTION

Consider a completely general set of canonical equations

$$\frac{dq_j}{dt} = \frac{\partial F}{\partial p_j}; \quad \frac{dp_j}{dt} = -\frac{\partial F}{\partial q_j} \quad 1 \leqslant j \leqslant N \tag{21}$$

and a change of variables $(q_j, p_j) \to (Q_j, P_j)$ defined as follows. We are given an arbitrary function S of $2N$ variables; it is called a *determining function* and is written as a function of the new variables Q_j and the old variables $p_j : S(Q_j, p_j); 1 \leq j \leq N$.

The change of variables is defined by the following $2N$ implicit equations

$$q_j = \frac{\partial S}{\partial p_j}, \qquad P_j = \frac{\partial S}{\partial Q_j} \tag{22}$$

We shall show that for any S the change of variables is canonical and does not change the characteristic function.

We evaluate the quantity

$$E = \sum_j (P_j \, dQ_j - p_j \, dq_j)$$

Differentiation of $S(Q_i, p_i)$ gives the identity

$$dS = \sum_j \frac{\partial S}{\partial Q_j} dQ_j + \sum_j \frac{\partial S}{\partial p_j} dp_j$$

or, from the definition of P_j and q_j in (22):

$$dS = \sum_j P_j \, dQ_j + \sum_j q_j \, dp_j \tag{23}$$

We then have

$$E = dS - \sum_j q_j \, dp_j - \sum_j p_j \, dq_j = d\left[S - \sum_j p_j q_j\right]$$

This is a total differential. Condition (17) is satisfied by $K=0$ (invariance of the Hamiltonian) and $W = S - \sum p_j q_j$.

B. CONJUGATE VARIABLES TO Q_j

Consider a system with canonical variables x_j and y_j. Suppose that we wish to make a canonical transformation which leaves the Hamiltonian unchanged and such that the q_j are given functions of the x_i. Relation (17) gives

$$\sum_j y_j \, dx_j - \sum_j p_j \, dq_j = dW \tag{24}$$

As the x_j are functions of the q_j, we have

$$dx_j = \sum_i \frac{\partial x_j}{\partial q_i} dq_i \qquad i, j = 1, 2, ..., n$$

If we put $dW = 0$ *a priori*, (24) is identically satisfied if

$$p_i = \sum_j y_j \frac{\partial x_j}{\partial q_i} \qquad i, j = 1, 2, ..., n \tag{25}$$

Now, if T has the form of section 19 with respect to the x_j

$$T = \frac{1}{2}\sum_j m_j \left(\frac{dx_j}{dt}\right)^2 = \frac{1}{2}\sum_j m_j \left(\sum_i \frac{\partial x_j}{\partial q_i}\frac{dq_i}{dt}\right)^2$$

we have, with $q_i' = (dq_i/dt)$,

$$\frac{\partial T}{\partial q_i'} = \sum_j m_j \frac{\partial x_j}{\partial q_i}\frac{dx_j}{dt} = \sum_j y_j \frac{\partial x_j}{\partial q_i}$$

Consequently, comparison with the right hand-side of (25) shows that

$$p_i = \frac{\partial T}{\partial q_i'}$$

where T has been written as a function of the q_i and the q_i'.

24. Jacobi's Theorem

We shall now establish a theorem which will subsequently allow us to define a very important system of canonical variables for the two-body problem.

We seek a canonical transformation which makes the new Hamiltonian zero.

For the canonical system:

$$\frac{dq_j}{dt} = \frac{\partial F}{\partial p_j}; \quad \frac{dp_j}{dt} = -\frac{\partial F}{\partial q_j} \quad 1 \leqslant j \leqslant N$$

we make the canonical transformation $(p_j, q_j) \rightarrow (P_j, Q_j)$ such that

$$\sum_j P_j \, dQ_j - \sum_j p_j \, dq_j + F \, dt = -dW \tag{26}$$

(we write $-dW$ to simplify future notation). W is a function of t, since F is non-zero. We equate the coefficients of the $(N+1)$ differentials, term by term:

$$P_j = -\frac{\partial W}{\partial Q_j}; \quad p_j = +\frac{\partial W}{\partial q_j}; \quad F(q_j, p_j, t) = -\frac{\partial W}{\partial t} \tag{27}$$

Replacing p_j by $\partial W/\partial q_j$ in the last equation, we obtain

$$F\left(q_j, \frac{\partial W}{\partial q_j}, t\right) + \frac{\partial W}{\partial t} = 0 \tag{28}$$

which is Jacobi's equation.

Suppose that the change of variables has been made. The new Hamiltonian is $F - F = 0$. The equations become

$$\frac{dP_j}{dt} = 0; \quad \frac{dQ_j}{dt} = 0$$

or

$$P_j = \beta_j; \quad Q_j = \alpha_j$$

where α_j and β_j are constants. The problem is thus solved if we can find this particular change of variables.

Assume that we have found a solution of (28) which depends on N linearly independent arbitrary constants a_j, on the N q_j, and on t:

$$W(q_j, a_j, t) = 0.$$

We make the canonical transformation such that the new variables Q_j allow the numbers a_j as solutions. The conjugate variables P_j are defined by the first series of equations (27)

$$P_j = b_j = - \frac{\partial W(q_j, a_j, t)}{\partial a_j} \tag{29}$$

The b_j are the constant values taken by P_j in the solution. The N relations (29) allow determination of the N variables q_j as functions of the $2N$ integration constants a_j and b_j, and of t; this holds, if each q_j effectively appears in each of the N equations. This condition precludes that one of the a_j, for example a_N, is an additive constant, i.e., that W is of the form $W(q_j, a_j, ..., a_{N-1}, t) + a_N$. We then substitute the values of q_j in the second of the series of equations in (27), $p_j = \partial W(q_j, a_j, t)/\partial q_j$, thus obtaining the N variables p_j as a function of a_j, b_j, and t.

Jacobi's theorem can be summarized as follows:

To integrate the system of $2N$ canonical equations

$$\frac{dq_j}{dt} = \frac{\partial F}{\partial p_j}; \quad \frac{dp_j}{dt} = -\frac{\partial F}{\partial q_j}$$

we find a complete solution of the Jacobi equation

$$F\left(q_j, \frac{\partial W}{\partial q_j}, t\right) + \frac{\partial W}{\partial t} = 0$$

This solution depends on N linearly independent arbitrary constants a_j. We then solve for q_j and p_j the system of equations

$$b_j = - \frac{\partial W(q_j, a_j, t)}{\partial a_j}; \quad p_j = \frac{\partial W(q_j, a_j, t)}{\partial q_j}$$

where the N quantities b_j are the missing N integration constants.

NOTE:

We can introduce one of the integration constants, say a_N, into Jacobi's equation by writing the latter as:

$$F\left(q_j, \frac{\partial W}{\partial q_j}, t\right) + \frac{\partial W}{\partial t} = a_N$$

This amounts to adding $a_N \, dt$ to both sides of (26), which evidently does not change the condition except that the Hamiltonian is no longer zero; it is equal to a_N, i.e. Q_N.

25. Canonical Equations for the Two-Body Problem

We shall apply Jacobi's theorem to the elliptic case of the two body problem. In fact, we have already found an elementary solution in Chapter II. The present treatment is complementary in that it will indicate which of the orbital elements form a system of canonical conjugate variables.

We can apply the results of section 18 to the two-body case; the equations of motion of one of the bodies are given by the system (10) after division of m_j

$$\frac{dx_j}{dt} = \frac{\partial F}{\partial y_j}; \quad \frac{dy_j}{dt} = -\frac{\partial F}{\partial x_j} \qquad j = 1, 2, 3$$

where x_1, x_2, x_3 are the Cartesian coordinates of the body, and

$$F = T - V = \tfrac{1}{2}(y_1^2 + y_2^2 + y_3^2) - \frac{\mu}{r}$$

keeping only the first term of V.

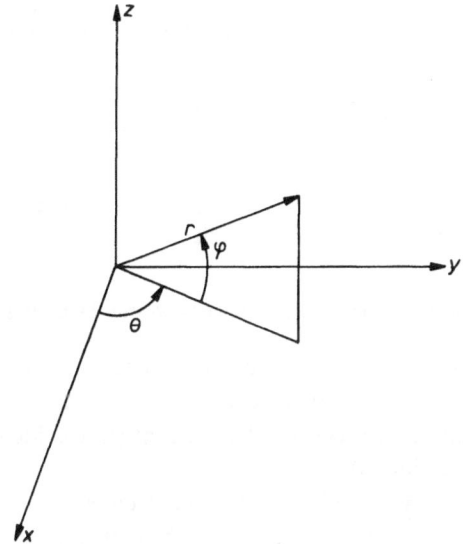

Fig. 5.

We write these equations in spherical polar coordinates (see Figure 5):

$$x = r \cos \varphi \cos \theta$$
$$y = r \cos \varphi \sin \theta$$
$$z = r \sin \varphi$$

F will now be expressed in these coordinates. It is easily shown that:

$$\left(\frac{dx}{dt}\right)^2 + \left(\frac{dy}{dt}\right)^2 + \left(\frac{dz}{dt}\right)^2 = \left(\frac{dr}{dt}\right)^2 + r^2\left(\frac{d\varphi}{dt}\right)^2 + r^2 \cos^2 \varphi \left(\frac{d\theta}{dt}\right)^2$$

We take the variables

$$q_1 = r \quad q_2 = \varphi \quad q_3 = \theta \tag{30}$$

and let q_1', q_2' and q_3' be their derivatives

$$\frac{dr}{dt}, \frac{d\varphi}{dt} \quad \text{and} \quad \frac{d\theta}{dt}$$

In this notation, we have

$$T = \tfrac{1}{2}q_1'^2 + \tfrac{1}{2}q_1^2 q_2'^2 + \tfrac{1}{2}q_1^2\, q_3'^2 \cos^2 q_2$$

Let us find the variables conjugate to the q_i. According to the result at the end of section 23, these are

$$p_1 = \frac{\partial T}{\partial q_1'} = q_1'; \quad p_2 = \frac{\partial T}{\partial q_2'} = q_1^2 q_2';$$

$$p_3 = \frac{\partial T}{\partial q_3'} = q_1^2\, q_3' \cos^2 q_2 \tag{31}$$

The characteristic function F becomes in these new variables:

$$F = \tfrac{1}{2}p_1^2 + \frac{1}{2q^2}\, p_2^2 + \frac{1}{2q_1^2 \cos^2 q_2}\, p_3^2 - \frac{\mu}{q_1}$$

and the equations are:

$$\frac{dq_i}{dt} = \frac{\partial F}{\partial p_i}; \quad \frac{dp_i}{dt} = -\frac{\partial F}{\partial q_i} \quad i = 1, 2, 3$$

26. Application of Jacobi's Theorem to the Two-Body Problem

We apply Jacobi's theorem to this problem, using the characteristic function F written above. Since F does not depend on t, $F = h$ is an integral of the system of equations; h is a constant (see section 21). This integral is identical with the kinetic energy integral (5) of section 10.

F can be replaced everywhere by $F - h$ without changing the equations. We shall also substitute $F - h$ for F in Jacobi's equation, thus introducing an arbitrary constant at the outset (see the Note to Section 24.)

Jacobi's equation is then:

$$\frac{1}{2}\left(\frac{\partial W}{\partial q_1}\right)^2 + \frac{1}{2q_1^2}\left(\frac{\partial W}{\partial q_2}\right)^2 + \frac{1}{2q_1^2 \cos^2 q_2}\left(\frac{\partial W}{\partial q_3}\right)^2 - \frac{\mu}{q_1} - h = 0 \tag{32}$$

Note that we do not want a complete solution of this equation, but simply a solution depending on three arbitrary constants. As the derivatives are separable, we can look for a solution W whose variables are also separable; i.e. we seek W in the form:

$$W = W_1(q_1) + W_2(q_2) + W_3(q_3) \tag{33}$$

Jacobi's equation is then

$$\frac{1}{2}\left(\frac{dW_1}{dq_1}\right)^2 + \frac{1}{2q_1^2}\left(\frac{dW_2}{dq_2}\right)^2 + \frac{1}{2q_1^2\cos^2 q_2}\left(\frac{dW_3}{dq_3}\right)^2 - \frac{\mu}{q_1} - h = 0$$

As the three bracketed quantities are independent, the equation is satisfied if, *for example*, we take the three following equations:

$$\frac{dW_3}{dq_3} = a_3$$

$$\frac{1}{2}\left(\frac{dW_2}{dq_2}\right)^2 + \frac{a_3^2}{2\cos^2 q_2} = \frac{a_2^2}{2}$$

$$\frac{1}{2}\left(\frac{dW_1}{dq_1}\right)^2 + \frac{a_2^2}{2q_1^2} - \frac{\mu}{q_1} - h = 0$$

Direct substitution shows that these make the left-hand side of Jacobi's equation identical with zero. We then have from (33):

$$W = \int \left(2h + \frac{2\mu}{q_1} - \frac{a_2^2}{q_1^2}\right)^{1/2} dq_1 + \int \left(a_2^2 - \frac{a_3^2}{\cos^2 q_2}\right)^{1/2} dq_2 + \int a_3 \, dq_3 \qquad (34)$$

which is a solution of (32) up to an additive constant due to the indefinite integrals; this solution depends on the three arbitrary constants a_3, a_2, and h, which we call a_1. Note that there is no need to define the signs of the square roots at this stage.

27. Meaning of the Constants a

We have shown in section 24 that the integration constants a_1, a_2, and a_3 appearing in W are the values assumed by the new variables Q_1, Q_2, and Q_3 of a system of canonical equations equivalent to the initial system, but whose characteristic function is identically zero. The solution is of the form:

$$Q_1 = a_1; \quad Q_2 = a_2; \quad Q_3 = a_3.$$

Let us find the meaning of these three canonical variables in elliptic motion.
(1) a_1 is what we call the kinetic energy constant h. From section 12:

$$h = \frac{-\mu}{2a}$$

where a is the semi-major axis.
(2) Consider the basic equation for a canonical transformation:

$$\sum_i P_i \, dQ_i - \sum_i p_i \, dq_i + F \, dt = -dW \qquad (35)$$

From (34), W depends on q_3 only through:

$$\int a_3 \, dq_3 = a_3 q_3$$

The only term of dW in dq_3 will thus be $a_3 \, dq_3$. Identifying the terms in dq_3 from (35), we have

$$- p_3 = - a_3$$

or

$$a_3 = q_1^2 \cos^2 q_2 q_3' = r^2 \cos^2 \varphi \frac{d\theta}{dt}$$

which is the z-component of the angular momentum. The magnitude of this angular momentum is, from formula (14) of section 12:

$$C = na^2 \sqrt{1 - e^2} = \sqrt{\mu a (1 - e^2)}$$

Its projection on Oz is $\sqrt{\mu a (1 - e^2)} \cos i$, and we have:

$$a_3 = Q_3 = \sqrt{\mu a (1 - e^2)} \cos i \, .$$

(3) We can similarly identify $-p_2$ with the coefficient of dq_2 in the total differential of dW. Here again, only q_2 and dq_2 appear in dW_2, and consequently:

$$- p_2 = - \sqrt{a_2^2 - \frac{a_3^2}{\cos^2 q_2}} = - \sqrt{a_2^2 - \frac{a_3^2}{\cos^2 \varphi}}$$

Replacing a_3 by the quantity $r^2 \cos^2 \varphi (d\theta/dt)$ found above, we have

$$p_2 = \sqrt{a_2^2 - r^4 \left(\frac{d\theta}{dt}\right)^2 \cos^2 \varphi}$$

then

$$p_2 = q_1^2 q_2' = r^2 \frac{d\varphi}{dt}$$

whence

$$a_2^2 = r^4 \left(\left(\frac{d\varphi}{dt}\right)^2 + \left(\frac{d\theta}{dt}\right)^2 \cos^2 \varphi\right)$$

This is the square of the angular momentum of the body about O.
Consequently $Q_2 = a_2 = \sqrt{\mu a (1 - e^2)}$.

28. Variables Conjugate to the Q_i

According to Jacobi's theorem, the variables P_j assume the constant values defined by $b_j = -\partial W / \partial a_j$ where W is given by (34). However, we shall fix this function W, because as written, it is defined only up to an arbitrary constant and the signs of the integrals. Another choice would give a different, equally valid, system of variables P_j.

We take:

$$W = \int_{q_1(t_0)}^{q_1(t)} \varepsilon_1 \left(2a_1 + \frac{2\mu}{q_1} - \frac{a_2^2}{q_1^2} \right)^{1/2} dq_1 + \int_0^{\varphi} \varepsilon_2 \left(a_2^2 - \frac{a_3^2}{\cos^2 q_2} \right)^{1/2} dq_2 + \int_0^{\theta} a_3 \, dq_3$$

where t_0 is the instant of perifocus passage; $\varepsilon_1 = +1$ if $q_1 = r$ is increasing and $\varepsilon_1 = -1$ if r is decreasing. This definition makes the derivative of the function under the first integral continuous, it being zero for values of q corresponding to passages through the apofocus and perifocus (see (1) below). ε_2 is similarly chosen so that the quantity under the second integration sign has a continuous derivative. Discontinuity occurs when $q_2 = \varphi = i$, so we shall take ε_2 as $+1$ when φ is increasing or when the distance $\omega + v = \psi$ to the node is between $-\pi/2$ and $+\pi/2$ (Fig. 6), hence $\cos \psi > 0$. Similarly, $\varepsilon_2 = -1$ if $\cos \psi < 0$.

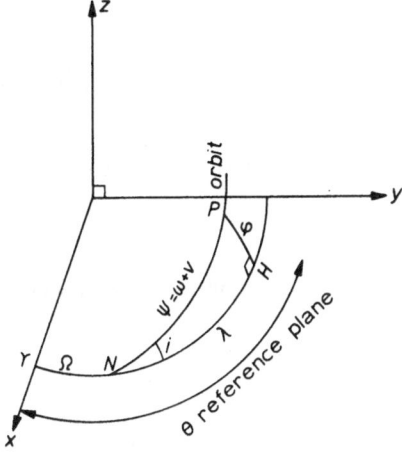

Fig. 6.

(1) The variable conjugate to Q_1 is given by

$$P_1 = -\frac{\partial W}{\partial a_1} = -\int_{q_1(t_0)}^{q_1(t)} \varepsilon_1 \left(2a_1 + \frac{2\mu}{q_1} - \frac{a_2^2}{q_1^2} \right)^{-1/2} dq_1$$

Now, q_1 is r; $a_1 = h = -\mu/2a$ and $a_2^2 = \mu a(1 - e^2)$. Multiplying above and below by $r(>0)$, we consequently have:

$$P_1 = -\varepsilon_1 \int_{r(t_0)}^{r(t)} \frac{r \, dr}{\sqrt{-(\mu/a)r^2 + 2\mu r - \mu a(1 - e^2)}}$$

To effect the integration, we express these quantities in terms of the eccentric anomaly E, for which we know that $r = a(1 - e \cos E)$; $dr = ae \sin E \, dE$ (see formula (12) of Chapter II).

Let E be the eccentric anomaly at t. By hypothesis, it is zero at t_0

$$P_1 = -\int_0^E \frac{\varepsilon_1 a(1 - e\cos E)\, ae\sin E\, dE}{\sqrt{\mu a}\,[-(1 - e\cos E)^2 + 2(1 - e\cos E) - 1 + e^2]^{1/2}}$$

$$P_1 = -\varepsilon_1 \int_0^E \frac{a^2(1 - e\cos E)\, e\sin E\, dE}{\sqrt{\mu a}\, e\,|\sin E|}$$

From the definition of ε_1 we have that

$$\varepsilon_1 \frac{\sin E}{|\sin E|} = +1.$$

Thus:

$$\left(2a_1 + \frac{2\mu}{q_1} - \frac{a_2^2}{q_1^2}\right)^{1/2} = \frac{|\sin E|}{r}$$

is zero on passage through the perifocus, as mentioned above. It follows that:

$$P_1 = -\int_0^E \frac{a\sqrt{a}}{\sqrt{\mu}}(1 - e\cos E)\, dE = \frac{-1}{n}(E - e\sin E) = -(t - t_0) \cdot$$

according to Kepler's law.

The final Hamiltonian is $h = Q_1$, using the note in section 24.

The equation giving P_1 is:

$$\frac{dP_1}{dt} = -\frac{\partial F}{\partial Q_1} = -1$$

which integrates to give $P_1 = -t + b_1$.

The constant of integration b_1 is thus t_0, the instant of passage through the perihelion.

(2) Similar calculations give $P_3 = -\Omega$ and $P_2 = -\omega$, where we must take the same precautions in defining the signs and the end points of integration. Ω and ω are respectively the longitude of the ascending node and the argument of the perigee.

29. Application of these Results to the General Problem

We have thus established a new system of conjugate variables:

$$\left.\begin{array}{l} Q_1 = -\dfrac{\mu}{2a}; \quad Q_2 = \sqrt{\mu a(1 - e^2)}; \quad Q_3 = \sqrt{\mu a(1 - e^2)}\cos i \\[2mm] P_1 = -t + t_0; \quad P_2 = -\omega; \quad P_3 = -\Omega \end{array}\right\} \tag{36}$$

whose characteristic function for the two-body problem reduces to $Q_1 = -\mu/2a$.

We could make this Hamiltonian zero by a suitable canonical transformation, but such new transformation is not desirable for the rest of the calculation.

We now consider the general case, such as that arising in the three-body problem (section 17). Let us consider only the equations for one of the bodies; it is clear that the following equations can be extended to the other.

The system of equations we had arrived at was (equation 10):

$$\frac{dx_j}{dt} = \frac{\partial F}{\partial y_j}; \quad \frac{dy_j}{dt} = \frac{\partial F}{\partial x_j} \quad j = 1, 2, 3 \tag{37}$$

F is in the form of $F = T - V$ and, as we have seen, V contains the term μ/r. We put

$$V = \frac{\mu}{r} + R$$

R is called the *disturbing function*. In the Hamiltonian $F = (T - \mu/r) - R$, R accounts for all that is not a two-body problem in the motion studied.

$F^* = T - \mu/r$ is the Hamiltonian of the two-body problem which we have just discussed.

Equations (37) become

$$\frac{dx_j}{dt} = \frac{\partial (F^* - R)}{\partial y_j}; \quad \frac{dy_j}{dt} = \frac{-\partial (F^* - R)}{dx_j}$$

Let us now transform the variables of this system so that the *new variables* are the quantities P_1, P_2, P_3, Q_1, Q_2, Q_3 defined by equations (36). System (37) is not identical with the two-body system of section 27 and, as it has a different solution, the new variables will no longer be constants in the solution. We say that they are the *osculating* quantities in the study of motion; they represent the elements of motion for a two-body case in which the only force is the central force μ/r. The osculating elements are of course defined in the set of axes, x_1, x_2, x_3 relative to the body in question.

This change of variables is such that the new Hamiltonian F_1 is:

$$F_1 = F - \left(F^* + \frac{\mu}{2a} \right) = -R - \frac{\mu}{2a}$$

since this transformation has reduced the Hamiltonian F^* to $Q_1 = -\mu/2a$.

Consequently, the system (37) is equivalent to the system:

$$\frac{dQ_j}{dt} = \frac{\partial (-R - (\mu/2a))}{\partial P_j}; \quad \frac{dP_j}{dt} = -\frac{\partial (-R - (\mu/2a))}{\partial Q_j} \quad j = 1, 2, 3 \tag{38}$$

We can improve the appearance of this system by changing all the signs of the Hamiltonian, and by putting $P'_j = -P_j$:

$$P'_1 = t - t_0; \quad P'_2 = \omega \quad P'_3 = \Omega \tag{39}$$

We then have the following system:

$$\frac{dQ_j}{dt} = +\frac{\partial (R + (\mu/2a))}{\partial P'_j}; \quad \frac{dP'_j}{dt} = -\frac{\partial (R + (\mu/2a))}{\partial Q_j} \tag{40}$$

30. The Delaunay Variables

The variables P_2' and P_3' now represent two classical elliptical elements. We can attempt to transform P_1' in a manner such that it represents the mean anomaly. We denote by $L, G, H, l, g,$ and h, the six new canonical variables which would be obtained after this transformation, trying to affect this transformation in such a manner that the characteristic function remains unchanged, as well as P_2' and P_3' which must be equal to g and h respectively.

The condition for this transformation to be canonical and for the Hamiltonian to remain unchanged is

$$l\, dL + g\, dG + h\, dH - P_1'\, dQ_1 - P_2'\, dQ_2 - P_3'\, dQ_3 = dW$$

It is desired that:

$$P_2' = g; \quad P_3' = h \quad \text{and} \quad l = n(t - t_0) = nP_1' = \sqrt{\mu}\, a^{-3/2} P_1'$$

These conditions will be fulfilled if

$$Q_2 = G, \quad Q_3 = H$$

and

$$l\, dL - P_1'\, dQ_1 = dW$$

$$\sqrt{\mu}\, a^{-3/2} P_1'\, dL - P_1'\, d\left(\frac{-\mu}{2a}\right) = dW$$

$$P_1'\left(\sqrt{\mu}\, a^{-3/2}\, dL - \frac{\mu\, da}{2a^2}\right) = dW$$

A possible solution is $dW = 0$, whence

$$\frac{\sqrt{\mu}\, da}{2\sqrt{a}} = dL$$

so that

$$L = \sqrt{\mu a}$$

Consequently, if we put again

$$\Phi = \frac{\mu}{2a} + R = \frac{\mu^2}{2L^2} + R$$

the system of equations given by (37) is equivalent to the system

$$\left. \begin{array}{lll} \dfrac{dL}{dt} = \dfrac{\partial\Phi}{\partial l}; & \dfrac{dG}{dt} = \dfrac{\partial\Phi}{\partial g}; & \dfrac{dH}{dt} = \dfrac{\partial\Phi}{\partial h} \\[2mm] \dfrac{dl}{dt} = -\dfrac{\partial\Phi}{\partial L}; & \dfrac{dg}{dt} = -\dfrac{\partial\Phi}{\partial G}; & \dfrac{dh}{dt} = -\dfrac{\partial\Phi}{\partial H} \end{array} \right\} \tag{41}$$

Thus we suppose that Φ is expressed as a function of the variables $L, G, H, l, g,$ and h

whose significance with regard to the elliptical elements is

$$L = \sqrt{\mu a}; \quad G = \sqrt{\mu a (1 - e^2)}; \quad H = \sqrt{\mu a (1 - e^2)} \cos i \left.\begin{array}{l}\\\\\end{array}\right\}$$
$$l = M = n(t - t_0); \quad g = \omega; \quad h = \Omega \qquad\qquad\qquad \tag{42}$$

This system of canonical variables, called Delaunay variables, was of great importance in the development of the theory of the Moon and remains one of the most effective systems used in perturbation problems. We shall use it particularly in Chapter V.

31. Osculating Elements

We have seen in section 8 that, in general, elliptical motion constitutes a correct approximation to the real motion observed in the solar system. Thus for example if, starting from an instant t_0, all the perturbing forces were neglected, the movement of a body would become exactly elliptical. It would represent the real movement quite well for a certain time, even though strictly speaking, it would not be identical with the real movement as regards position and velocity except at the instant t_0. The elements of an ellipse that would be followed by a body after a specific time t are said to be *osculating* or *instantaneous* if, starting from this instant, all the forces with the exception of the central force were to disappear. The elements of such an unperturbed orbit can be defined at any instant: they correspond to the elliptical orbit followed by a moving body which would have at the given instant the same position and the same velocity as the real body.

As in fact the real orbit is simply tangential to the osculating orbit* at an instant $t + \delta t$ the osculating orbit will be different, with different osculating elements. It follows that the osculating elements in perturbed motion are no longer constant but *functions of time*.

The same reasoning can be applied to Delaunay's variables. When the perturbations disappear at an instant t, Φ becomes $\mu/2a (R=0)$ and the solution of the equations are L, G, H, g, h (constants), $l = n(t - t_0)$. Thus we see that in the general case Delaunay's variables are also osculating variables in the sense given above. They are connected with elliptical osculating elements by the formulae (42).

Osculating elements are often used to describe the perturbed motion of a body. They possess the advantage of having a precise and simple geometrical significance (see section 13 in Chapter II) while having small variations.

If, however, one wants to pass from these osculating elements to rectangular coordinates, or vice versa, the transformation is easy. Thus, let us consider the osculating elements at the instant t. By definition, the position and velocity of the body at instant t are such that the motion of the two bodies defined by these quantities has the osculating elements as elements of the ellipse. We shall therefore have the coordinates and the velocity at the instant t in computing the coordinates and the velocity at the instant t corresponding to the osculating elements at the same instant t. The formu-

---

* Note that the osculating orbit, contrary to the geometrical significance of this term, is only tangential and not "osculating" with the real orbit.

lation in section 14 therefore remains valid, and the following important result is obtained.

The coordinates and velocity components of perturbed motion at an instant t are those which would be obtained at this instant t, assuming that the orbit is elliptical, from elements equal to the osculating elements at the same instant t.

32. Lagrange Equations

In view of the importance of osculating elements as variables in celestial mechanics, we shall establish the differential equations equivalent to the systems already given, but where the variables are elliptical osculating elements.

We start from the Delaunay equations given by (41) with the six variables L, G, H, l, g, and h, and effect a change of variables defined by the relations (42), written in a differential form

$$
\left.
\begin{aligned}
dL &= \frac{\sqrt{\mu}}{2\sqrt{a}}\, da \\[2mm]
dG &= \frac{\sqrt{\mu}\sqrt{1-e^2}}{2\sqrt{a}}\, da - \frac{\sqrt{\mu a}\, e}{\sqrt{1-e^2}}\, de \\[2mm]
dH &= \frac{\sqrt{\mu}\sqrt{1-e^2}\cos i}{2\sqrt{a}}\, da - \frac{\sqrt{\mu a}\, e \cos i}{\sqrt{1-e^2}}\, de - \sqrt{\mu a(1-e^2)}\sin i\, di
\end{aligned}
\right\} \tag{43}
$$

$$
dl = dM; \quad dg = d\omega \quad \text{and} \quad dh = d\Omega \tag{44}
$$

From (43) with the aid of relations (41) and noting that $\Phi = \mu/2a + R$, we obtain

$$
\frac{da}{dt} = \frac{2\sqrt{a}}{\sqrt{\mu}}\frac{dL}{dt} = \frac{2\sqrt{a}}{na^{3/2}}\frac{\partial\Phi}{\partial l} = \frac{2}{na}\frac{\partial R}{\partial M}
$$

$$
\frac{de}{dt} = \frac{-\sqrt{1-e^2}}{\sqrt{\mu a}\, e}\frac{dG}{dt} + \frac{\sqrt{\mu}\,(1-e^2)}{2\sqrt{a}\sqrt{\mu a}\, e}\frac{da}{dt} = \frac{-\sqrt{1-e^2}}{na^2 e}\frac{\partial R}{\partial\omega} + \frac{1-e^2}{na^2 e}\frac{\partial R}{\partial M}
$$

$$
\frac{di}{dt} = \frac{-1}{\sqrt{\mu a}\sqrt{1-e^2}\sin i}\frac{dH}{dt} + \frac{\sqrt{\mu}\sqrt{1-e^2}\cos i}{2\sqrt{a}\sqrt{\mu a}\sqrt{1-e^2}\sin i}\frac{da}{dt}
$$

$$
- \frac{\sqrt{\mu a}\, e \cos i}{\sqrt{1-e^2}\sqrt{\mu a}\sqrt{1-e^2}\sin i}\frac{de}{dt}
$$

$$
= \frac{-1}{na^2\sqrt{1-e^2}\sin i}\frac{\partial R}{\partial\Omega} + \frac{\cos i}{na^2\sqrt{1-e^2}\sin i}\frac{\partial R}{\partial\omega}
$$

On the other hand, rearrangement of (42) gives

$$
a = \frac{L^2}{\mu}; \quad \sqrt{1-e^2} = \frac{G}{L}
$$

whence $e=\sqrt{1-(G^2/L^2)}$ and finally $\cos i=H/G$.

The three differential equations (44) give

$$\frac{dh}{dt}=\frac{d\Omega}{dt}=-\frac{\partial R}{\partial H}=-\frac{\partial R}{\partial i}\frac{\partial i}{\partial H}=\frac{1}{na^2\sqrt{1-e^2}\sin i}\frac{\partial R}{\partial i}$$

$$\frac{dg}{dt}=\frac{d\omega}{dt}=-\frac{\partial R}{\partial G}=-\frac{\partial R}{\partial e}\frac{\partial e}{\partial G}-\frac{\partial R}{\partial i}\frac{\partial i}{\partial G}$$

$$=-\frac{\partial R}{\partial e}\times\frac{-G}{L^2}\times\frac{1}{\sqrt{1-(G^2/L^2)}}-\frac{\partial R}{\partial i}\times\frac{-1}{\sin i}\times\frac{-H}{G^2}$$

$$=\frac{\sqrt{1-e^2}}{na^2e}\frac{\partial R}{\partial e}-\frac{\cos i}{na^2\sqrt{1-e^2}\sin i}\frac{\partial R}{\partial i}$$

$$\frac{dl}{dt}=\frac{dM}{dt}=-\frac{\partial}{\partial L}\left(\frac{\mu}{2a}\right)-\frac{\partial R}{\partial L}$$

$$=-\frac{\partial}{\partial L}\left(\frac{\mu^2}{2L^2}\right)-\frac{\partial R}{\partial a}\frac{\partial a}{\partial L}-\frac{\partial R}{\partial e}\frac{\partial e}{\partial L}$$

$$=\frac{\mu^2}{L^3}-\frac{\partial R}{\partial a}\times\frac{2L}{\mu}-\frac{\partial R}{\partial e}\times\frac{G^2}{L^3}\frac{1}{\sqrt{1-(G^2/L^2)}}$$

$$=n-\frac{2}{na}\frac{\partial R}{\partial a}-\frac{1-e^2}{na^2e}\frac{\partial R}{\partial e}$$

The system of equations obtained in this way, equivalent to the Delaunay system, constitutes the Lagrange equations given below:

$$\left.\begin{array}{l}
\dfrac{da}{dt}=\dfrac{2}{na}\dfrac{\partial R}{\partial M}\\[2ex]
\dfrac{de}{dt}=\dfrac{-\sqrt{1-e^2}}{na^2e}\dfrac{\partial R}{\partial\omega}+\dfrac{1-e^2}{na^2e}\dfrac{\partial R}{\partial M}\\[2ex]
\dfrac{di}{dt}=\dfrac{-1}{na^2\sqrt{1-e^2}\sin i}\dfrac{\partial R}{\partial\Omega}+\dfrac{\cos i}{na^2\sqrt{1-e^2}\sin i}\dfrac{\partial R}{\partial\omega}\\[2ex]
\dfrac{d\Omega}{dt}=\dfrac{1}{na^2\sqrt{1-e^2}\sin i}\dfrac{\partial R}{\partial i}\\[2ex]
\dfrac{d\omega}{dt}=\dfrac{\sqrt{1-e^2}}{na^2e}\dfrac{\partial R}{\partial e}-\dfrac{\cos i}{na^2\sqrt{1-e^2}\sin i}\dfrac{\partial R}{\partial i}\\[2ex]
\dfrac{dM}{dt}=n-\dfrac{2}{na}\dfrac{\partial R}{\partial a}-\dfrac{1-e^2}{na^2e}\dfrac{\partial R}{\partial e}
\end{array}\right\} \quad (45)$$

NOTE:

Just as in the preceding sections, in these equations n represents merely $\sqrt{\mu}/a^{3/2}$.

It is no longer a constant since *a* is no longer a constant. In particular, in the sixth of these equations the *n* term will be obtained with the same approximation in relation to the small quantities to be found in *R* as the two other terms, after integration of the first equation. We must therefore carry out a double integration of the equation giving the semi-major axis before being able to obtain the mean anomaly. This is an important general result. Whatever method is used, no problem of perturbed trajectory can be solved in celestial mechanics without carrying out a double integration at some stage. This result will have important consequences in specifying the results, particularly when long-period terms or numerical integrations are involved.

33. The Case of Zero Eccentricity or Zero Inclination

The Lagrange formulae are no longer valid when *e* or *i* are zero because these quantities appear in the denominators in some of these equations. Moreover, we shall see when solving a problem of celestial mechanics in Chapter V with the help of a system of Delaunay canonical equations, that this system also leads to impossible results for small *e* or small *i*. This is due to the choice of osculating elements as variables.

A. SMALL ECCENTRICITY

Let us consider an ellipse of very slight eccentricity subject to perturbations of the order to this eccentricity, and suppose that these perturbations tend to shorten the major axis *AA'* and elongate the minor axis *BB'*. In other words, the eccentricity will diminish. The osculating ellipse will be deformed in a continuous manner until *BB'* becomes greater than *AA'*. The eccentricity will then again begin to increase, after passing through the value $e=0$. However, at the same time the argument of the perigee will have changed by 90°, its origin being thus sharply changed. The anomaly will be subject to a discontinuity of the same quantity. The solution, put into the form of variation of osculating elements, will therefore be discontinuous. On the other hand, $\omega + M$ remains a continuous variable, with the origin remaining fixed or varying continuously.

We can thus find other variables which remain continuous when *e* passes through zero. These are:

$$\eta = e \sin \omega; \quad \theta = e \cos \omega; \quad \lambda = \omega + M \tag{46}$$

The change of variables thus defined can be effected in the equations given by (45). We shall not write down the formulation obtained which is seldom used.

B. SMALL INCLINATION

If the perturbations are such that the plane of the osculating orbit can turn round the axis of the nodes on traversing the reference plane, the situation presents some analogy with the previous one. The inclination passes through zero, remaining positive while the line of the nodes remains unchanged, the significance of the two nodes is reversed, and the longitude of the ascending node changes sharply by 180°.

The origin of the argument of the periastron being thus suddenly changed, the latter is also subject to a discontinuity of 180°.

We are obliged once again to change the variables in a manner analogous to the case of zero eccentricity. In general, we put:

$$p = \tan i \sin \Omega; \quad q = \tan i \cos \Omega; \quad \varpi = \Omega + \omega \qquad (47)$$

CHAPTER IV

PERTURBATION THEORY

34. Introduction

In the preceding chapter we have given systems of differential equations whose variables were osculating elliptical elements or were closely related to the latter (Delaunay variables).

However, the right-hand sides of these equations were functions of the disturbing function R, the expression for which was given as a function of the rectangular coordinates of the moving body. We therefore find an annoying heterogeneity. It is necessary that the two sides are expressed as a function of the same variables. We shall devote the first part of this chapter to expressing R as a function of the osculating elements, and having obtained the form of the equations we shall try to find the form of the solution, giving a possible procedure for its construction. However, before tackling these problems, let us consider some analytical results which will be particularly useful.

35. Fourier Series

Let us recall the following result concerning trigonometric series: if $f(x)$ is a bounded periodic function of period 2π, with a bounded variation whatever the value of x, and integrable, the product of $f(x)$ and $\cos nx$ or $\sin nx$ (n whole) is also integrable.

If

$$a_0 = \frac{1}{2\pi} \int_0^{2\pi} f(t)\, dt; \qquad a_p = \frac{1}{\pi} \int_0^{2\pi} f(t) \cos pt\, dt$$

$$b_p = \frac{1}{\pi} \int_0^{2\pi} f(t) \sin pt\, dt \tag{1}$$

the series

$$a_0 + \sum_{p=1}^{\infty} (a_p \cos px + b_p \sin px) \tag{2}$$

is the Fourier expansion of $f(x)$. With the conditions given above, Jordan's theorem enables us to confirm that it converges for all the values of x, its sum being:

$$\frac{f(x+0) + f(x-0)}{2}$$

In particular, if the function $f(x)$ is continuous the Fourier series converges uniformly whatever the value of x, and its sum is $f(x)$.

36. Expansion of the Eccentric Anomaly in Fourier Series

We have seen that the eccentric anomaly E and the mean anomaly M are connected by the implicit equation known as Kepler's equation:

$$E - e \sin E = M \tag{3}$$

(formula 18 of Chapter II)

If $e < 1$, $dE/dM = 1/(1 - e \cos E)$ is periodic with a period of 2π and continuous. From the result in the preceding section we can therefore expand E into a uniformly converging Fourier series. Thus, let us expand the quantity

$$E - M = e \sin E$$

We apply the formulae (1):

$$a_0 = \frac{1}{2\pi} \int_0^{2\pi} e \sin E \, dM$$

with

$$dM = (1 - e \cos E) \, dE$$

We take E as the variable: integration limits remain unchanged. We find

$$a_0 = \frac{1}{2\pi} \int_0^{2\pi} e \sin E \, dE - \frac{1}{2\pi} \int_0^{2\pi} e^2 \sin E \cos E \, dE = 0.$$

In a similar way, one would get

$$a_p = \frac{1}{\pi} \int_0^{2\pi} e \sin E \cos pM \, dM = \frac{1}{\pi} \int_{-\pi}^{+\pi} e \sin E \cos pM \, dM = 0,$$

since it is the integral from $-\pi$ to $+\pi$ of an odd function. Finally,

$$\pi b_p = \int_0^{2\pi} e \sin E \sin pM \, dM = \int_0^{2\pi} e \sin E (1 - e \cos E) \sin p \left[E - e \sin E \right] dE$$

We integrate by parts, putting

$$du = \sin pM \, dM;$$

$$u = \frac{-1}{p} \cos pM = -\frac{1}{p} \cos p \, (E - e \sin E)$$

$$v = e \sin E; \qquad dv = e \cos E \, dE$$

$$\pi b_p = \left[-\frac{1}{p} e \sin E \cos p \, (E - e \sin E) \right]_0^{2\pi} + \int_0^{2\pi} \frac{1}{p} \cos p \, (E - e \sin E) \, e \cos E \, dE$$

The part entirely integrated is zero. On the other hand, applying the formula

$$\cos a \cos b = \tfrac{1}{2} \left[\cos (a + b) + \cos (a - b) \right],$$

we obtain

$$\pi b_p = \frac{e}{2p} \int_0^{2\pi} \cos \left[(p + 1) E - pe \sin E \right] dE + \frac{e}{2p} \int_0^{2\pi} \cos \left[(p - 1) E - pe \sin E \right] dE$$

We introduce the function

$$J_k(x) = \frac{1}{2\pi} \int_0^{2\pi} \cos (kt - x \sin t) \, dt \tag{4}$$

The expansion of $E - M$ in Fourier series is written

$$E = M + e \sum_{p=1}^{\infty} \frac{J_{p-1}(pe) + J_{p+1}(pe)}{p} \sin pM \tag{5}$$

37. Definition of Bessel Functions

Let a function Z of the complex variable z be:

$$Z = e^{\frac{x}{2}\left(z - \frac{1}{z}\right)}$$

where x is a real number

$$Z = e^{\frac{xz}{2}} \, e^{-\frac{x}{2z}}$$

Each factor can be expanded into an absolutely converging series in $xz/2$ or $x/2z$ if $|z| \neq 0$.

We have

$$e^{\frac{xz}{2}} = \sum_{m=0}^{\infty} \left(\frac{x}{2}\right)^m \frac{1}{m!} z^m; \quad e^{-\frac{x}{2z}} = \sum_{n=0}^{\infty} \left(-\frac{x}{2}\right)^n \frac{1}{n!} z^{-n}$$

The product is an absolutely converging double series

$$Z = \sum_{m=0}^{\infty} \sum_{n=0}^{\infty} \frac{(-1)^n}{m! \, n!} \left(\frac{x}{2}\right)^{m+n} z^{m-n} \tag{6}$$

We can regroup the terms of the same order in z, $m-n=k$, and write

$$Z = \sum_{k=-\infty}^{+\infty} J_k(x) \, z^k \tag{7}$$

with

$$J_0(x) = 1 - \left(\frac{x}{2}\right)^2 \frac{1}{(1!)^2} + \left(\frac{x}{2}\right)^4 \frac{1}{(2!)^2} - \cdots + \left(\frac{x}{2}\right)^{2n} \frac{(-1)^n}{(n!)^2} + \cdots \tag{8}$$

$$J_k(x) = \left(\frac{x}{2}\right)^k \frac{1}{k!} \left[1 - \left(\frac{x}{2}\right)^2 \frac{1}{1!(k+1)} + \cdots \right.$$

$$\left. + \left(\frac{x}{2}\right)^{2n} \frac{(-1)^n}{n!(k+1)(k+2)\ldots(k+n)} + \cdots \right] \tag{9}$$

$J_k(x)$ is called a Bessel function of k-th order, and can be calculated from expansions (8) and (9).

Let us put

$$z = e^{it}; \quad z - \frac{1}{z} = e^{it} - e^{-it} = 2i \sin t$$

whence

$$Z = e^{ix \sin t} = \sum_{-\infty}^{+\infty} J_k(x) \, e^{ikt}$$

This is simply a Fourier series put in the form of imaginary exponentials. Let us calculate the integral

$$\int_0^{2\pi} e^{ix \sin t} \, e^{-ijt} \, dt = \sum_{-\infty}^{+\infty} J_k(x) \int_0^{2\pi} e^{(k-j)it} \, dt \tag{10}$$

now,

$$\int_0^{2\pi} e^{i\lambda t} \, dt = \int_0^{2\pi} (\cos \lambda t + i \sin \lambda t) \, dt = \begin{cases} 0 & \text{if } \lambda \neq 0 \\ 2\pi & \text{if } \lambda = 0 \end{cases}$$

The right-hand side of (10) has only one term which is non-zero, obtained for $j=k$; we have therefore

$$2\pi J_k(x) = \int_0^{2\pi} e^{ix\sin t} e^{-ikt}\, dt = \int_0^{2\pi} e^{-i(kt-x\sin t)}\, dt$$

$$= \int_0^{2\pi} \cos(kt - x\sin t)\, dt - i\int_0^{2\pi} \sin(kt - x\sin t)\, dt$$

Since the second integral is zero (odd function of the period 2π):

$$J_k(x) = \frac{1}{2\pi} \int_0^{2\pi} \cos(kt - x\sin t)\, dt \tag{11}$$

which is identical with expression (4). The coefficients introduced in series (5) are the expressions dependent on Bessel's functions such as were just defined, which can be calculated from (8) or (9).

38. Some Properties of Bessel Functions

(a) It is easy to verify in formulae (9) that

$$J_{-k}(x) = (-1)^k J_k(x)$$
$$J_{-k}(-x) = J_k(x)$$

If x is small $J_k(x)$ is of the order of x^k, the principal term being

$$\frac{1}{k!}\left(\frac{x}{2}\right)^k$$

(b) We have

$$e^{\frac{x}{2}\left(z-\frac{1}{z}\right)} = \sum_{-\infty}^{+\infty} J_k(x)\, z^k$$

Differentiating with respect to z:

$$\frac{x}{2}\left(1 + \frac{1}{z^2}\right)\sum_{-\infty}^{+\infty} J_k(x)\, z^k = \sum_{-\infty}^{+\infty} k J_k(x)\, z^{k-1}$$

Identifying the coefficients of z^{k-1} on both sides, we obtain

$$J_k(x) = \frac{x}{2k}\left[J_{k-1}(x) + J_{k+1}(x)\right] \tag{12}$$

(c) However, if we differentiate with respect to x:

$$\frac{1}{2}\left(z - \frac{1}{z}\right) \sum_{-\infty}^{+\infty} J_k(x) \, z^k = \sum_{-\infty}^{+\infty} \frac{\mathrm{d}J_k(x)}{\mathrm{d}x} \, z^k$$

Identification of terms in z^k gives:

$$\frac{\mathrm{d}J_k(x)}{\mathrm{d}x} = \tfrac{1}{2}[J_{k-1}(x) - J_{k+1}(x)] \tag{13}$$

(d) Differentiating once again with respect to x:

$$\frac{\mathrm{d}^2 J_k(x)}{\mathrm{d}x^2} = \frac{1}{2}\left[\frac{\mathrm{d}J_{k-1}(x)}{\mathrm{d}x} - \frac{\mathrm{d}J_{k+1}(x)}{\mathrm{d}x}\right] = \frac{1}{4}[J_{k-2}(x) - 2J_k(x) + J_{k+2}(x)]$$

Applying formulae (12) and (13) we obtain

$$\frac{\mathrm{d}^2 J_k(x)}{\mathrm{d}x^2} + \frac{1}{x}\frac{\mathrm{d}J_k(x)}{\mathrm{d}x} + \left(1 - \frac{k^2}{x^2}\right)J_k(x) = 0 \tag{14}$$

This is the differential equation whose solutions are Bessel functions of k-th order.

(e) Consider the two series:

$$e^{\frac{x}{2}\left(z - \frac{1}{z}\right)} = J_0(x) + \sum_1^\infty J_k(x) \, z^k + \sum_1^\infty (-1)^k \, J_k(x) \, z^{-k}$$

$$e^{-\frac{x}{2}\left(z - \frac{1}{z}\right)} = J_0(x) + \sum_1^\infty J_k(x) \, z^{-k} + \sum_1^\infty (-1)^k \, J_k(x) \, z^k$$

By multiplying these two expressions term by term we obtain:

$$1 = J_0^2(x) + 2J_1^2(x) + 2J_2^2(x) + \cdots$$

We have deduced the important inequalities

$$|J_0(x)| \leqslant 1; \quad |J_k(x)| \leqslant \frac{1}{\sqrt{2}}$$

39. Expansion of $\cos jE$ and $\sin jE$

With the aid of formula (12) we can write down the result of the example in section 36 in the following simple manner

$$E = M + \sum_{p=1}^\infty \frac{2J_p(pe)}{p} \sin pM \tag{15}$$

A calculation analogous to that described in this example allows us to expand in the same manner $\cos jE$ and $\sin jE$. Thus, if we put:

$$\cos jE = a_0^{(j)} + \sum_{p=1}^{\infty} a_p^{(j)} \cos pM + \sum_{p=1}^{\infty} b_p^{(j)} \sin pM$$

$$dM = (1 - e \cos E)\, dE$$

we have

$$a_0^{(j)} = \frac{1}{2\pi} \int_0^{2\pi} \cos jE\, dM = \frac{1}{2\pi} \int_0^{2\pi} \cos jE\,(1 - e \cos E)\, dE$$

which gives

$$a_0^{(1)} = -\frac{e}{2}; \quad a_0^{(j>1)} = 0$$

$$a_p^{(j)} = \frac{1}{\pi} \int_0^{2\pi} \cos jE \cos pM\, dM = \frac{1}{p\pi} \int_0^{2\pi} \cos jE\, \frac{d(\sin pM)}{dM}\, dM$$

Let us integrate by parts

$$p\pi a_p^{(j)} = \left[\sin jE \sin pM \right]_0^{2\pi} - \int_0^{2\pi} \sin pM\, \frac{d(\cos jE)}{dM}\, dM$$

$$= +j \int_0^{2\pi} \sin jE \sin p\,(E - e \sin E)\, dE$$

$$= \frac{j}{2} \int_0^{2\pi} \cos\left[(p - j)E - pe \sin E \right] dE - \frac{j}{2} \int_0^{2\pi} \cos\left[(p + j)E - pe \sin E \right] dE$$

$$a_p^{(j)} = \frac{j}{p} \left[J_{p-j}(pe) - J_{p+j}(pe) \right]$$

It may be shown that $b_p^{(j)} = 0$ using the fact that we have odd periodic functions. We thus find

$$\cos jE = \left\{ \begin{array}{l} -\dfrac{e}{2}(j=1) \\[4pt] \\ 0 \quad j \neq 1 \end{array} \right\} + \sum_{p=1}^{\infty} \frac{j}{p} \left[J_{p-j}(pe) - J_{p+j}(pe) \right] \cos pM \qquad (16)$$

We similarly have

$$\sin jE = \sum_{p=1}^{\infty} \frac{j}{p} \left[J_{p-j}(pe) + J_{p+j}(pe) \right] \sin pM \qquad (17)$$

40. Expression of Other Functions of the Two-Body Problem

Starting from these expansions we can easily obtain those of other quantities which appear in the two-body problem. Let us consider a few examples.

(a) $r/a = 1 - e \cos E$. Formula (16) gives the expansion of $\cos E$, whence, with formula (13)

$$\frac{r}{a} = 1 + \frac{e^2}{2} - \sum_{p=1}^{\infty} \frac{2e}{p^2} \frac{dJ_p(pe)}{de} \cos pM \tag{18}$$

(b) $(r/a)^k = (1 - e \cos E)^k$ can be written in the form $\sum_{j=0}^{k} \alpha_j \cos jE$, the α_j being functions of e.

Formula (16) can again be applied. For example

$$\left(\frac{r}{a}\right)^3 = (1 - e \cos E)^3 = 1 - 3e \cos E + 3e^2 \cos^2 E - e^3 \cos^3 E$$

$$= 1 + \frac{3e^2}{2} - \left(3e + \frac{3e^3}{4}\right) \cos E + \frac{3e^2}{2} \cos 2E - \frac{e^3}{4} \cos 3E$$

With the aid of (16), we have

$$\left(\frac{r}{a}\right)^3 = 1 + 3e^2 + \frac{3e^4}{8} + \sum_{p=1}^{\infty} \left[\frac{-12e - 3e^3}{4p} (J_{p-1}(pe) - J_{p+1}(pe)) \right.$$

$$\left. + \frac{3e^2}{p} (J_{p-2}(pe) - J_{p+2}(pe)) - \frac{3e^3}{4p} (J_{p-3}(pe) - J_{p+3}(pe)) \right] \cos pM \tag{19}$$

(c)

$$\frac{a}{r} = \frac{1}{1 - e \cos E} = \frac{dE}{dM} = 1 + \sum_{p=1}^{\infty} 2J_p(pe) \cos pM \tag{20}$$

(d) From the equation of the ellipse

$$r = \frac{a(1 - e^2)}{1 + e \cos v}$$

we have

$$\cos v = -\frac{1}{e} + \frac{1 - e^2}{e} \times \frac{a}{r} = -e + \sum_{p=1}^{\infty} 2 \frac{(1 - e^2)}{e} J_p(pe) \cos pM \tag{21}$$

On the other hand, $r/a = (1 - e^2)/(1 + e \cos v)$ gives on differentiation

$$\frac{d}{dM}\left(\frac{r}{a}\right) = \frac{1}{1 - e^2} \left(\frac{r}{a}\right)^2 e \sin v \frac{dv}{dM}$$

Thus from the law of areas

$$r^2 \frac{dv}{dM} = \frac{C}{n} = a^2 \sqrt{1 - e^2}$$

whence

$$\frac{d}{dM}\left(\frac{r}{a}\right) = \frac{e \sin v}{\sqrt{1 - e^2}}$$

$$\sin v = \frac{\sqrt{1 - e^2}}{e} \frac{d}{dM}\left(\frac{r}{a}\right) = \sqrt{1 - e^2} \sum_{p=1}^{\infty} \frac{2}{p} \frac{dJ_p(pe)}{de} \sin pM \qquad (22)$$

(e) The reduced coordinates of a point on the ellipse, in relation to the major axis and to the perpendicular axis passing through the focus are [see formula (11), Chapter II]:

$$x = r \cos v = a\,(\cos E - e) = a\left[-\frac{3e}{2} + \sum_{p=1}^{\infty} \frac{1}{p}(J_{p-1}(pe) - J_{p+1}(pe)) \cos pM \right] \quad (23)$$

$$y = r \sin v = a\sqrt{1 - e^2}\,\sin E$$

$$= a\sqrt{1 - e^2}\left[\sum_{p=1}^{\infty} \frac{1}{p}(J_{p-1}(pe) + J_{p+1}(pe)) \sin pM \right] \qquad (24)$$

(f) The differential equations of the motion in x and y are

$$\frac{d^2x}{dt^2} = -\mu \frac{x}{r^3}; \quad \frac{d^2y}{dt^2} = -\mu \frac{y}{r^3}$$

Thus

$$\frac{x}{r^3} = -\frac{1}{\mu}\frac{d^2x}{dM^2}\left(\frac{d^2M}{dt^2}\right) = -\frac{1}{\mu}\frac{\mu}{a^3}\left(\frac{d^2x}{dM^2}\right)$$

Differentiating the preceding equations twice with respect to M, we obtain

$$\frac{x}{r^3} = \frac{1}{a^2} \sum_{p=1}^{\infty} p\,[J_{p+1}(pe) - J_{p-1}(pe)] \cos pM \qquad (25)$$

$$\frac{y}{r^3} = \frac{-\sqrt{1 - e^2}}{a^2} \sum_{p=1}^{\infty} p\,[J_{p+1}(pe) + J_{p-1}(pe)] \sin pM \qquad (26)$$

In this manner we could obtain the Fourier expansions of most functions involved in the two-body problem, either directly or by combining the preceding expansions (operations on the series).

41. Relation between E and v

We know the relation

$$\tan\frac{E}{2} = \sqrt{\frac{1-e}{1+e}}\,\tan\frac{v}{2}$$

(formula (13), Chapter II). We now put

$$p = \sqrt{\frac{1-e}{1+e}}; \quad \text{we have} \quad \tan\frac{E}{2} = p\tan\frac{v}{2} \tag{27}$$

Using imaginary exponentials, this can be written

$$\frac{e^{iE}-1}{e^{iE}+1} = p\,\frac{e^{iv}-1}{e^{iv}+1}$$

or alternatively

$$e^{iE} = \frac{e^{iv}(1+p)-(p-1)}{-e^{iv}(p-1)+(p+1)}$$

Putting further $q=(1-p)/(p+1)$ from (27)): $q=(1-\sqrt{1-e^2})/e)$ we have

$$e^{iE} = e^{iv}\,\frac{1+qe^{-iv}}{1+qe^{iv}}$$

Taking the logarithms:

$$iE = iv + \ln(1 + qe^{-iv}) - \ln(1 + qe^{iv})$$

If q, which is of the order of the eccentricity, is small, it is possible to expand the two log functions into series. Division by i then gives:

$$E = v + 2\left(q\,\frac{-e^{iv}+e^{-iv}}{2i} + \frac{q^2}{2}\,\frac{e^{2iv}-e^{-2iv}}{2i} + \cdots + (-1)^n\frac{q^n}{n}\,\frac{-e^{niv}+e^{-niv}}{2i} + \cdots\right)$$

$$E = v - 2q\sin v + \frac{2q^2}{2}\sin 2v - \cdots + (-1)^n\frac{2q^n}{n}\sin nv + \cdots \tag{28}$$

Similarly, by interchanging p and $1/p$, q becomes $-q$ and

$$v = E + 2q\sin E + \frac{2q^2}{2}\sin 2E + \cdots + \frac{2q^n}{n}\sin nE + \cdots \tag{29}$$

42. D'Alembert's Property

In most applications of celestial mechanics we are led to consider that the eccentricity e is small (this is in fact the case in the majority of bodies in the solar system). Under these conditions, property (a) of section 37 enables us to give a qualitative view of the aspect of the preceding series. Suppose we wish to find the predominant terms in the Fourier series and to see how the other terms behave.

In series (16) and (17), in $\cos jE$ and $\sin jE$, the most important term is that which contains $J_0(pe)$ which occurs when $p-j=0$ or $p=j$.

The principal term, which will be of zero order in e, is the term in $\cos jM$ or $\sin jM$. The "neighbouring" terms [in $(j-1)M$ or $(j+1)M$] will contain $J_1(pe)$ and $J_{-1}(pe)$. They will therefore be of the first order in e. Similarly, the term in $(j-n)M$ and $(j+n)M$ at a "distance" n, will be of the n-th order in e.

Finally, we see that the term in $\cos pM$, containing J_{p-j} and J_{p+j} where $p+j$ and $p-j$ are always of like parity, will be formed of terms of even or odd order, and that two neighbouring terms are of different parity.

It can also be seen, from (b) in section 40, that $(r/a)^k$ has an analogous property, the term of zero order being the constant term. It follows that the expansions of $(r/a)^k \cos jE$ and $(r/a)^k \sin jE$ will have the same principal terms as $\cos jE$ or $\sin jE$, and that the property concerning the order of neighbouring terms will be conserved.

It can further be shown, even though the calculation is more complex, that this also applies to $(r/a)^{-k}$. Finally, the relations between E and v given by (28) and (29) satisfy the same properties, with a term of zero order for v. The expansions of q'' are all either even or odd in e. It can be deduced from this that expansions of the type

$$\left(\frac{r}{a}\right)^{\pm k} \sin jv \quad \text{or} \quad \left(\frac{r}{a}\right)^{\pm k} \cos jv$$

will behave in the same manner as those of the analogous expressions in jE.

Finally, we shall describe the following property, called d'Alembert's property which can be checked, for example, on the series of section 40.

Expressions of the form

$$\left(\frac{r}{a}\right)^{\pm k} \sin jv; \quad \left(\frac{r}{a}\right)^{\pm k} \cos jv; \quad \left(\frac{r}{a}\right)^{\pm k} \sin jE;$$

$$\left(\frac{r}{a}\right)^{\pm k} \cos jE; \quad \left(\frac{r}{a}\right)^{\pm k} x^j; \quad \left(\frac{r}{a}\right)^{\pm k} y^j$$

can be represented in Fourier series.

The coefficients are odd or even series in e. The order in e of the coefficient of $\cos pM$ or $\sin pM$ is $|p-j|$.

This property enables us to say that if a term of a series is of zero order, the neighbouring terms are of order 1, the subsequent ones of order 2, etc. If the eccentricity e is small, it will be possible to keep only a certain number of terms on either side of the central term and to consider the others as negligible. Thus, in the expansion of x/r^3 (25), if we can consider e^4 to be negligible it is sufficient to keep the three terms following the principal term, which is in $\cos M$ (since here $j=1$). We shall thus keep the terms in $\cos M$, $\cos 2M$, $\cos 3M$, and $\cos 4M$, as well as the constant term.

43. Limited Expansions in e

The expansions of sections 39 and 40 are rather difficult to use owing to the presence of Bessel functions in the coefficients. However, just as we can neglect certain terms in Fourier series when the coefficient is of a high order in e, we can expand the Bessel functions into series in e [formula (9)], and limit these series at terms which we consider negligible. This is the generally adopted procedure.

For example, if we neglect e^4, some of the series in sections 40 and 41 become

$$\frac{r}{a} = 1 + \frac{e^2}{2} + (-e + \tfrac{3}{8}e^3)\cos M - \frac{e^2}{2}\cos 2M - \tfrac{3}{8}e^3\cos 3M \qquad (30)$$

$$\frac{a}{r} = 1 + \left(e - \frac{e^3}{8}\right)\cos M + e^2\cos 2M + \tfrac{9}{8}e^3\cos 3M \qquad (31)$$

$$x = -\frac{3e}{2} + (1 - \tfrac{3}{8}e^2)\cos M + \left(\frac{e}{2} - \frac{e^3}{3}\right)\cos 2M + \tfrac{3}{8}e^2\cos 3M + \frac{e^3}{3}\cos 4M \qquad (32)$$

$$y = (1 - \tfrac{5}{8}e^2)\sin M + \left(\frac{e}{2} - \tfrac{5}{12}e^3\right)\sin 2M + \tfrac{3}{8}e^2\sin 3M + \frac{e^3}{3}\sin 4M \qquad (33)$$

$$E = M + \left(e - \frac{e^3}{8}\right)\sin M + \frac{e^2}{2}\sin 2M + \tfrac{3}{8}e^3\sin 3M \qquad (34)$$

$$v = M + \left(2e - \frac{e^3}{4}\right)\sin M + \tfrac{5}{4}e^2\sin 2M + \tfrac{13}{12}e^3\sin 3M \qquad (35)$$

The last series is obtained by substituting the series (34) into (29).

44. Convergence of Series Expanded in Powers of e

We have already seen that the series of section 40 are absolutely convergent whatever the value of $e < 1$. However, when considered as double series in e and M, they are not absolutely convergent: a change in the order of terms modifies the radius of convergence. The convergence is maintained if e is small.

By way of example, it can be shown that the expression (34) for E as a function of M, written in the expanded form

$$E = M + e\sin M + \frac{e^2}{2}\sin 2M + e^3\left(-\tfrac{1}{8}\sin M + \tfrac{3}{8}\sin 3M\right) + \cdots \qquad (36)$$

converges only for $e < 0.6627,\ldots$

The same limit applies for the other series.

It should be noted that the finite expressions (30) to (35) limited to a certain power of e are in fact only the first terms of series in increasing powers of eccentricity. Thus the above limit of convergence is the practical limit of validity of the truncated expressions used in practice.

With the few exceptions (comets, some asteroids, Jupiter VIII), the instantaneous eccentricities of the bodies in the solar system never reach this limit. In the majority of cases, we have $e < 0.2$ if not $e < 0.1$, and the use of the approximate expressions is justified on the condition that enough terms are taken for the remaining ones to be effectively negligible. In choosing the terms to be kept for a limiting power of e, use will be made of d'Alembert's property.

45. Expression of the Disturbing Function (the Case of the Moon)

Let us now express the disturbing function in its entirety as a function of the osculating elements. We consider the disturbing function R of the three-body problem where the third body has a negligible mass. From the definition of R (section 29) and the expression for V_2 of section 18, we obtain:

$$R = km \left(\frac{1}{\Delta} - \frac{x'x + y'y + z'z}{r'^3} \right) \tag{37}$$

where x', y', z' $(x'^2 + y'^2 + z'^2 = r'^2)$ are coordinates of the perturbing body, moving on an ellipse, and Δ is the distance of this perturbing body from the body being studied, the latter having coordinates x, y, $z(r)$.

We have

$$\Delta^2 = (x - x')^2 + (y - y')^2 + (z - z')^2 = r'^2 - 2(x'x + y'y + z'z) + r^2$$

If we suppose for example that $r' \gg r$, which is the case of the Moon perturbed by the Sun, we have

$$\frac{1}{\Delta} = \frac{1}{r'} \left[1 - \frac{2(xx' + yy' + zz')}{r'^2} + \frac{r^2}{r'^2} \right]^{-1/2}$$

Designating the angle between lines joining the principal body to the two others by S, we have

$$\cos S = \frac{xx' + yy' + zz'}{r'r}$$

whence

$$\frac{1}{\Delta} = \frac{1}{r'} \left[1 - \frac{2r}{r'} \cos S + \frac{r^2}{r'^2} \right]^{-1/2} \tag{38}$$

Since r/r' is small, the part between square brackets can be expanded into a power series in r/r', for example with the aid of the binomial formula

$$(1 - \varepsilon)^{-1/2} = 1 + \frac{\varepsilon}{2} + \tfrac{3}{8} \varepsilon^2 + \tfrac{5}{16} \varepsilon^3 + \cdots$$

$$\frac{1}{\Delta} = \frac{1}{r'} \left[1 + \frac{r}{r'} \cos S + \frac{r^2}{r'^2} (-\tfrac{1}{2} + \tfrac{3}{2} \cos^2 S) + \frac{r^3}{r'^3} (-\tfrac{3}{2} \cos S + \tfrac{5}{2} \cos^3 S) + \cdots \right]$$

This can be reintroduced into R. However, it may be noted that the equations of motion depend only on the partial derivatives of R with respect to the elements of the body studied, and consequently we can neglect in R any additional term depending only on the elements of the perturbing celestial body, such as $1/r'$.

Observing also that the second term of R destroys the second term of $1/\Delta$, we see that it is sufficient to keep for R:

$$R = \frac{km}{r'}\left[\frac{r^2}{r'^2}\left(-\tfrac{1}{2} + \tfrac{3}{2}\cos^2 S\right) + \frac{r^3}{r'^3}\left(-\tfrac{3}{2}\cos S + \tfrac{5}{2}\cos^3 S\right) + \cdots\right] \qquad (39)$$

NOTE:

The quantities in parentheses, resulting from the expansion of the part in square brackets (38), are called Legendre polynomials:

$$\begin{aligned}
P_0(x) &= 1 &&; \quad P_1(x) = x \\
P_2(x) &= \tfrac{1}{2}(3x^2 - 1); && \quad P_3(x) = \tfrac{1}{2}(5x^3 - 3x) \\
P_4(x) &= \tfrac{1}{8}(35x^4 - 30x^2 + 3), && \quad \text{etc...}
\end{aligned}$$

It can be shown that

$$P_n(x) = \frac{1}{2^n n!}\frac{d^n}{dx^n}(x^2 - 1)^n$$

and that

$$R = \frac{km}{r'}\sum_{n=2}^{\infty}\left(\frac{r}{r'}\right)^n P_n(\cos S)$$

46. Reduction to the Variables of Elliptical Motion

Let us continue to consider the case of the motion of a satellite perturbed by the Sun. R is given by (39). To simplify the calculations we shall give only the first term in the explicit form, i.e.:

$$R_1 = \frac{km}{r'^3}r^2\left(-\tfrac{1}{2} + \tfrac{3}{2}\cos^2 S\right) \qquad (40)$$

Let S and L be respectively the geocentric directions of the Sun and of the satellite. Furthermore, let N be the ascending node of the orbit of L (Figure 7), and let Ω, ω, and v be the longitude of the node, the argument of perigee, and the true anomaly of the satellite respectively, the same notation with primes being used for the corresponding elements of the Sun, and the origin of the longitude being γ.

$$\widehat{\gamma S} = \omega' + v'; \quad \widehat{NL} = \omega + v; \quad \widehat{\gamma N} = \Omega$$

We put

$$\psi = \Omega + \omega + v; \quad \psi' = \omega' + v'$$

whence

$$\widehat{NL} = \psi - \Omega; \quad \widehat{SN} = \psi' - \Omega$$

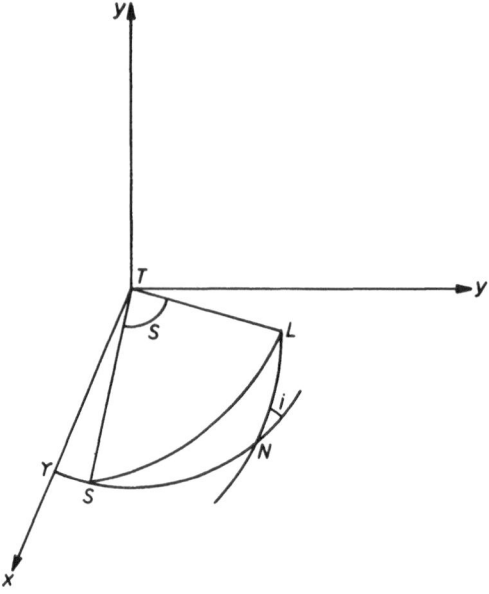

Fig. 7.

In the spherical triangle SNL, we have the following relations

$$\cos S = \cos(\psi - \Omega)\cos(\psi' - \Omega) + \sin(\psi - \Omega)\sin(\psi' - \Omega)\cos i$$

Replacing $\cos i$ by $\cos^2(i/2) - \sin^2(i/2)$ and multiplying the first term of the right-hand side by

$$\cos^2\frac{i}{2} + \sin^2\frac{i}{2},$$

we obtain

$$\cos S = \cos^2\frac{i}{2}\cos(\psi - \psi') + \sin^2\frac{i}{2}\cos(\psi + \psi' - 2\Omega)$$

whence

$$R_1 = \frac{kmr^2}{r'^3}\left[-\tfrac{1}{2} + \tfrac{3}{2}\cos^4\frac{i}{2}\cos^2(\psi - \psi')\right.$$

$$+ 3\cos^2\frac{i}{2}\sin^2\frac{i}{2}\cos(\psi - \psi')\cos(\psi + \psi' - 2\Omega)$$

$$\left. + \tfrac{3}{2}\sin^4\frac{i}{2}\cos^2(\psi + \psi' - 2\Omega)\right]$$

Returning to the true anomaly and to the other elements by

$$\psi - \psi' = \Omega + \omega + v - \omega' - v'$$

$$\psi + \psi' - 2\Omega = \omega + \omega' + v + v' - \Omega$$

$$km = n'^2a'^3$$

we see that it is possible to express R_1 – and in fact R – as a function of these elements:

$$R_1 = n'^2 a^2 \left(\frac{a'}{r'}\right)^3 \left(\frac{r}{a}\right)^2 \left[-\tfrac{1}{2} + \tfrac{3}{4} \cos^4 \frac{i}{2} \right.$$

$$+ \tfrac{3}{4} \cos^4 \frac{i}{2} \cos 2(\omega - \omega' + v - v' + \Omega)$$

$$+ \tfrac{3}{8} \sin^2 i \left[\cos(2\omega + 2v) + \cos(2\omega' + 2v' - 2\Omega) \right]$$

$$\left. + \tfrac{3}{4} \sin^4 \frac{i}{2} + \tfrac{3}{4} \sin^4 \frac{i}{2} \cos 2(\omega + \omega' + v + v' - \Omega) \right] \qquad (41)$$

47. Expansion of the Disturbing Function

Expressions of the type of (41) can be further transformed by separating v and v'. A simple but laborious calculation shows that the right-hand side of this equation can also be written as a sum of terms in the form

$$n'^2 a^2 \left(\frac{a'}{r'}\right)^3 \left(\frac{r}{a}\right)^2 f_1(v) f_2(v') f_3(i, \omega, \omega', \Omega) \qquad (42)$$

where $f_1(x)$ and $f_2(x)$ are one of the following functions: $\sin 2x$, $\cos 2x$, or 1, whilst f_3 is a more complicated trigonometric function.

Each of the following terms in (42)

$$\left(\frac{a'}{r'}\right)^3, \quad \left(\frac{a'}{r'}\right)^3 \cos 2v', \quad \left(\frac{a'}{r'}\right)^3 \sin 2v'$$

is a known function of time, and can be expressed in terms of e' and the mean anomaly M' of the Sun, by the use of formulae of the type shown in section 40.

On the other hand, the other terms in (42), i.e.

$$\left(\frac{r}{a}\right)^2, \left(\frac{r}{a}\right)^2 \cos 2v \quad \text{and} \quad \left(\frac{r}{a}\right)^2 \sin 2v$$

express the mean anomaly M of the satellite in the form of a Fourier series (the coefficients depend only on the corresponding eccentricity). Substituting these expressions we can find R as a function of time via M' and the six osculating elements of the satellite: a, e, i, Ω, ω, and M.

These expressions converge in the same way as those in the previous sections (cf. section 40 ff.).

In the case of small eccentricities e and e' (which is the general case in the solar system), the resulting series converge and we can neglect all terms of orders in e and e' higher than a certain value determined by the required accuracy.

In the case of a perturbing body moving in an elliptical orbit around the perturbed body, the disturbing function R_1, or quite generally R, assumes the form

$$R = \Sigma A_{l_1 l_2 l_3 l_4 l_5}(a, e, i, a', e') \cos(l_1 M + l_2 M' + l_3 \omega + l_4 \Omega + l_5 \omega') \tag{43}$$

where $l_1, ..., l_5$ are zero or positive or negative integers. Regrouping the terms in M and M' with coefficients depending on ω, ω', and Ω leads to the disappearance of the sine terms in the final expressions. To verify this property, we note that the odd series in (43) combine only with the sine terms, and the even series combine only with the cosine terms. This is by virtue of the formulae of the sine and cosine products derived above.

Furthermore, although in this method the coefficients A are functions of a, e, and i, they can also be easily expressed in terms of other osculating variables, whenever this suits our purpose. Thus, by changing the variables in accordance with

$$L = \sqrt{\mu a} \quad G = \sqrt{\mu a (1 - e^2)} \quad H = \sqrt{\mu a (1 - e^2)} \cos i$$

we can express A as a function of the Delaunay variables L, G, and H.

48. Expansion in a Small Parameter

It is a further noteworthy property of the disturbing function expressed in terms of the osculating variables that we can expand it into a rapidly converging series in a small parameter [cf. (39)]. Thus, we can put α, and α' equal to constants close to the mean values of r and r', and put $\alpha/\alpha' = \varepsilon$, where ε is small, being about $1/400$ in the case of the Moon. We can then write R in the form

$$R = \frac{km}{r'} \left[\varepsilon^2 \left(\frac{r}{\alpha}\right)^2 \left(\frac{\alpha'}{r'}\right)^3 (-\tfrac{1}{2} + \tfrac{3}{2} \cos^2 S) \right.$$
$$\left. + \varepsilon^3 \left(\frac{r}{\alpha}\right)^3 \left(\frac{\alpha'}{r'}\right)^3 (-\tfrac{3}{2} \cos S + \tfrac{5}{2} \cos^3 S) + \cdots \right]$$

where α'/α and r/α are close to 1, and R is thus expanded in a series that rapidly converges in ε.

In the case of planetary perturbations, where r and r' are of the same order of magnitude, the perturbation function depends on the planetary masses; these are very small in comparison with the mass of the Sun (see Chapter VII). We shall see in Chapter V that the disturbing function brought into play by the imperfect spherical shape of a planet can also be expanded into a rapidly converging series in terms of small parameters.

Clearly, just like the differentials, the solution will also depend on one or more small parameters. The usual practice has always been to expand the solution in terms of the same small parameters, and to neglect in the result terms beyond a certain exponent. However, it was Poincaré who first showed that this method is in fact justifiable. We shall now describe this result without its derivation.

49. Theorems of Proof

Cauchy has proved the existence of a holomorphic* solution of a system of differential equations which may be written in the simplified form as follows:

$$\frac{dx}{dt} = \varphi(x, y, t); \quad \frac{dy}{dt} = \psi(x, y, t)$$

If φ and ψ are holomorphic functions bounded in x and y and continuous in t for $|x| < R$, $|y| < R'$, and $t \leq r$, we can find solutions $x(t)$ and $y(t)$ that are holomorphic for $|t| < t_0 < r$.

As regards the set of differential equations

$$\frac{dx}{dt} = \varphi(x, y, \varepsilon, t); \quad \frac{dy}{dt} = \psi(x, y, \varepsilon, t)$$

where ε is a variable parameter close to zero, Poincaré's theorem, which is a generalized form of Cauchy's theorem, proves that φ and ψ (continuous in t and holomorphic in x and y) are also holomorphic in ε. Consider a certain solution

$$x(t, \varepsilon); \quad y(t, \varepsilon)$$

which is reduced to $x_0(t)$ and $y_0(t)$ when $\varepsilon = 0$. This solution can be expanded into complete series in ε:

$$\left.\begin{array}{l} x(t, \varepsilon) = x_0(t) + \varepsilon x_1(t) + \varepsilon^2 x_2(t) + \cdots \\ y(t, \varepsilon) = y_0(t) + \varepsilon y_1(t) + \varepsilon^2 y_2(t) + \cdots \end{array}\right\} \tag{44}$$

These series converge when $\varepsilon < \varepsilon_0 \neq 0$ irrespective of the value of t in the region in which the hypotheses are to be verified.

It follows that, since they satisfy the conditions of holomorphy, those differential equations used in celestial mechanics which depend on one or more small parameters (cf. section 48), have solutions that can be expanded in these small parameters.

In particular, since the disturbing function R and its derivatives can be considered to be small in comparison with the right-hand sides of the differential equations of motion in rectangular coordinates (eq. 13 in Chapter III), the general solution can be expanded in the small quantity characterizing R. Furthermore, if $R \to 0$, the solution is reduced to the solution in the case without a disturbing force, i.e. to the solution of the two-body problem.

50. Form of Equations with the Osculating Elements

The equations can be put in two groups (a) and (b) whether they are of type (41) with Delaunay variables or of type (45) with elliptical elements:

---

* A function is said to be holomorphic in a certain number of variables in a given region when it can be expanded into power series in these variables in the given region.

(a) Equations with metric variables a, e, and i, or L, G, and H (denoted in a general manner by L_i). The right-hand sides of these equations contain R in the form of partial differential derivatives with respect to the angular variables (incorporated in the cosine of R), as transformed in section 47 (eq. 43). These equations can be put in the general form:

$$\frac{dL_i}{dt} = \Sigma B_i(L_k) \sin(\Sigma \alpha_j l_j) \tag{45}$$

where the B_i terms depend on the metric variables L_j, the α_j terms are integers, and the l_j terms are angular variables or linear functions of time (if R depends on the position of other bodies, expressed e.g. as a function of the mean anomaly). The B_i terms contain the small parameter ε characteristic of R.

(b) Equations with angular variables l, g, and h, or M, ω, and Ω (denoted in a general manner by l_j). The right-hand sides of these equations contain R in the form of partial differential derivatives with respect to the metric variables in the coefficients of the cosine terms. Furthermore, some of these equations contain a term that depends only on metric variables without the small parameter as a factor. In the Lagrange equation in dM/dt, this term is $n = \sqrt{\mu} a^{-3/2}$, whilst in the dl/dt equation of the Delaunay system it is $\partial/\partial L(\mu^2/2L^2) = -\mu^2/L^3$, which is derived from the second term of $\Phi = R + \mu^2/2L^2$.

These equations can be put in the more general form of

$$\frac{dl_i}{dt} = a_i(L_k) + \Sigma b_{ij}(L_k) \cos(\Sigma \alpha_j l_j) \tag{46}$$

where a_i and b_i depend on the metric variables L_k; α_j and l_j have the same meaning as above. The a_i terms can be small (of the order of ε) or they can have a finite value, whilst all the b_{ij} terms contain the small parameter ε as a factor.

51. Method of Solution

The method we shall describe here is not the simplest, but it reveals the salient features of the solution.

According to Poincaré's theorem, if R is a function of ε (assumed to be small), the solution is in the form (44), which can be written as

$$\left. \begin{array}{l} L_i = L_{i_0} + \varepsilon L_{i_1} + \varepsilon^2 L_{i_2} + \cdots \\ l_i = l_{i_0} + \varepsilon l_{i_1} + \varepsilon^2 l_{i_2} + \cdots \end{array} \right\} \tag{47}$$

As mentioned before, L_{i_0} and l_{i_0} represent the solution of the two-body problem. These are constants or linear functions of time (for l and M).

Substitution of these expressions in (45) and (46) gives

$$\frac{dL_{i_0}}{dt} + \varepsilon \frac{dL_{i_1}}{dt} + \varepsilon^2 \frac{dL_{i_2}}{dt} + \cdots = \Sigma \left[\varepsilon B_{ij_1}(L_k) + \varepsilon^2 B_{ij_2}(L_k) + \cdots \right] \sin(\Sigma \alpha_j l_j)$$

$$\frac{dl_{i_0}}{dt} + \varepsilon \frac{dl_{i_1}}{dt} + \varepsilon^2 \frac{dl_{i_2}}{dt} + \cdots = a_i(L_k) + \Sigma \left[\varepsilon b_{ij_1}(L_k) + \varepsilon^2 b_{ij_2}(L_k) + \cdots \right] \cos(\Sigma \alpha_j l_j)$$

$$(48)$$

Having substituted the solution (47) we equate these equations in ε.

(a) Terms which are independent of ε:

$$\frac{dL_{i_0}}{dt} = 0; \quad \frac{dl_{i_0}}{dt} = a_i(L_{k_0}).$$

This shows that L_{i_0} is constant and l_{i_0} is a linear function of time:

$$l_{i_0} = a_i(L_{k_0})(t - t_0) = n_{i_0}(t - t_0)$$

(b) The terms in ε are derived from direct substitution of L_{k_0} into B_{ij_1} and b_{ij_1}, or from the terms in L_{i_0}. We thus have

$$\frac{dL_{i_1}}{dt} = \Sigma B_{ij_1}(L'_{k_0}) \sin(\Sigma \alpha_j l_{j_0})$$

$$\frac{dl_{i_1}}{dt} = \Sigma \frac{\partial a_i(L_{k_0})}{\partial L_{k_0}} L_{k_1} + \Sigma b_{ij_1}(L_{k_0}) \cos(\Sigma \alpha_j l_{j_0})$$

The equations in metric variables give

$$L_{i_1} = \frac{\Sigma B_{ij_1}(L_{k_0}) \cos(\Sigma \alpha_j l_{j_0})}{- \Sigma \alpha_j n_{j_0}} \qquad (49)$$

The n_{j_0} terms have already been determined. We need not consider the integration constant for L_{i_1}, since it has been introduced in L_{i_0} during the identification of terms independent of ε. We can now substitute (49) into the right-hand sides of the equations in dl_{i_1}/dt. The resulting series is of the form of

$$\frac{dl_{i_1}}{dt} = a_{i_1} + \Sigma b_{ij_1}(L_{k_0}) \cos(\Sigma \alpha_j l_{j_0})$$

Integration of this term by term gives

$$l_{i_1} = a_{i_1}(t - t_0) + \frac{\Sigma b'_{ij_1}(L_{k_0}) \sin(\Sigma \alpha_j l_{j_0})}{\Sigma \alpha_j n_{j_0}} \qquad (50)$$

We see that generally all angular variables will (but metric variables will not) contain a secular term, in which time in a linear form is outside the trigonometric functions.

(c) We can again substitute into the right-hand side of (49) the complete solution thus obtained and proceed to equate terms in ε^2. The calculation now becomes more complex. Let us start it for dL_{i2}/dt; we have

$$\frac{dL_{i_0}}{dt} + \varepsilon \frac{dL_{i_1}}{dt} + \varepsilon^2 \frac{dL_{i_2}}{dt} = \Sigma \left[\varepsilon B_{ij_1}(L_{k_0} + \varepsilon L_{k_1}) + \varepsilon^2 B_{ij_2}(L_{k_0})\right] \sin\left[\Sigma \alpha_j (\overline{l_{j_0}} + \varepsilon l_{j_1})\right]$$

$$= \Sigma \left[\varepsilon B_{ij_1}(L_{k_0}) + \varepsilon^2 \frac{\partial B_{ij_1}}{\partial L_{k_0}} L_{k_1} + \varepsilon^2 B_{ij_2}(L_{k_0})\right]$$

$$\times \left[\sin\left(\Sigma \alpha_j \overline{l_{j_0}}\right) \cos\left(\Sigma \alpha_j \varepsilon l_{j_1}\right) + \cos\left(\Sigma \alpha_j \overline{l_{j_0}}\right) \sin\left(\Sigma \alpha_j \varepsilon l_{j_1}\right)\right]$$

We expand l_j into $\overline{l_{j_0}} + \varepsilon l_{j_1}$, where the first term is the secular part of l_j and l_{j_1} is the periodic part containing ε as a factor. Identification of terms in ε^2 alone now gives

$$\cos \Sigma \alpha_j \varepsilon l_j = 1 \;(\text{up to } \varepsilon^2)$$

and

$$\sin \Sigma \alpha_j \varepsilon l_j = \alpha_j \varepsilon l_{j_1} \;(\text{up to } \varepsilon^3)$$

$$\frac{dL_{i_2}}{dt} = \Sigma B_{ij_1}(L_{k_0}) \varepsilon \alpha_j l_{j_1} \cos\left(\Sigma \alpha_j \overline{l_{j_0}}\right) + \Sigma \left[\frac{\partial B_{ij_1}}{\partial L_{k_0}} L_{k_1} + B_{ij_2}(L_{k_0})\right] \sin\left(\Sigma \alpha_j \overline{l_{j_0}}\right)$$

where

$$l_{i_1} = \Sigma B'_{ij_1}(L_{k_0}) \sin\left(\Sigma \alpha_j \overline{l_{j_0}}\right)$$

and:

$$L_{i_1} = \Sigma B''_{ij_1}(L_{k_0}) \cos\left(\Sigma \alpha_j \overline{l_{j_0}}\right)$$

The products of the form

$$\cos\left(\Sigma \alpha_j \overline{l_{j_0}}\right) \sin\left(\Sigma \alpha_i \overline{l_{j_0}}\right)$$

are transformed into terms in $\sin\left(\Sigma \alpha_k \overline{l_{k_0}}\right)$ with the aid of

$$\cos a \sin b = \tfrac{1}{2}\left(\sin(a+b) - \sin(a-b)\right)$$

Finally, dL_{i2}/dt assumes the form

$$\frac{dL_{i_2}}{dt} = \Sigma B_{ij_2} \sin\left(\Sigma \alpha_j \overline{l_{j_0}}\right)$$

Integration performed as before then gives an L_{i2} of the same form as L_{i1}.

The rest of the calculation is done as in point (b): this solution is substituted in a_{i0}, and the solution of (b) is substituted in the other terms. The products of the trigonometric functions obtained from the limited expansion of the right-hand side will now be in cosines. We then have that

$$\frac{dl_{i_2}}{dt} = a'_{i_2} + \Sigma b'_{ij_2}(L_{k_0}) \cos\left(\Sigma \alpha_j \overline{l_{j_0}}\right)$$

and l_{i2} is of the same form (eq. 50) as l_{i1}.

(d) The process can be continued ad infinitum, and the exponent of ε in the solution is increased by one in each repetition. The metric variables are always in the form of a cosine series:

$$L_i = L_{i_0} + \Sigma C_{ij}(L_{k_0}) \cos(\Sigma \alpha_j \bar{l}_{j_0}) \tag{51}$$

and the angular variables are in the form of a linear function of time and in the form of a sine series:

$$l_i = \bar{l}_{i_0} + \Sigma S_{ij}(L_{k_0}) \sin(\Sigma \alpha_j \bar{l}_{j_0}) \tag{52}$$

It should be added that the arguments of the trigonometric lines may contain not only the $\bar{l}_{j_0}$ secular terms of variables, but also other linear functions of time, which have already been encountered in the expansion of the disturbing function.

NOTE:

The important result to remember is that this process leads only to those linear secular terms which feature in the angular variables. In particular, there are no terms with $(t-t_0)^2$ as a factor, and no so-called mixed term in which $(t-t_0)$ would be included twice: once as a factor of the coefficient, and once in the arguments of the trigonometric line.

This result has featured in a theorem developed by Delaunay in the case of the Moon and was subsequently generalized by Tisserand to the case of the planets. It is of great importance in connection with the stability of any system that obeys these equations, though only when the convergence of the solution thus expressed is guaranteed. We shall see later that this condition is generally not fulfilled.

52. Long-Period and Short-Period Terms

We shall now apply the previous results to the theory of the motion of a satellite (the Moon) around a planet revolving around the Sun in accordance with Kepler's laws, this representing the principal problem in the theory of the Moon. We have seen in connection with the formation of the disturbing function in section 47 that R depends on five angular variables M, M', ω, Ω, and ω', where M, ω, and Ω are the osculating elements of the motion of the Moon, M' (mean anomaly of the Sun) is a linear function of time, and ω' is a constant angle (argument of the perigee of the Sun), which can be taken as zero if we choose correctly the origin of the longitudes. The metric variables of the problem, a, e, and i, correspond to the motion of the Moon. If a_0, e_0, and i_0 are the integration constants of a, e, and i, and if the secular parts of the angular variables Ω, ω, and M are given by

$$\bar{\Omega} = \Omega_0 + n_\Omega(t - t_0); \quad \bar{\omega} = \omega_0 = n_\omega(t - t_0)$$

and

$$\bar{M} = M_0 + n(t - t_0)$$

the solution can then be written as

$$a = a_0 + \sum_{\alpha\beta\gamma\delta} a_{\alpha\beta\gamma\delta}(a_0, e_0, i_0 \ldots) \cos(\alpha\bar{M} + \beta\bar{\omega} + \gamma\bar{\Omega} + \delta M') \tag{53}$$

The quantities e and i have an analogous form. We do not mention here the dependence of $a_{\alpha\beta\gamma\delta}$ on the orbital elements of the Sun. The expressions for the angular variables are exemplified by

$$\Omega = \bar{\Omega} + \sum_{\alpha\beta\gamma\delta} \Omega_{\alpha\beta\gamma\delta}(a_0, e_0, i_0 \ldots) \sin(\alpha\bar{M} + \beta\bar{\omega} + \gamma\bar{\Omega} + \delta M') \tag{54}$$

those for ω and M being analogous. In the summation, α, β, γ, and δ are integers or zero. They can represent any combination of values, but not all can be zero.

(a) We shall say that a term is a short-period term when α and δ are not zero at the same time. We then have the following possibilities concerning the movement of the Moon around the Earth:

$\alpha = 1$	$\delta =$ any value	the period is of the order of one sidereal revolution of the Moon (27 days)
$\alpha = 0$	$\delta = 1$	period $= 1$ year
$\alpha > 1$	$\delta =$ any value	period < 1 month
$\alpha = 0$	$\delta > 1$	period ≤ 6 months.

The time coefficients of $\bar{\omega}$ and $\bar{\Omega}$ are very small and hardly modify the above periods.

(b) We shall say that a term is a long-period term when $\alpha = \delta = 0$. Since the secular terms of ω and Ω appear only in the second approximation (terms in ε), n_ω and n_Ω are of the order of ε. The associated periods are thus of the order of the period of M (the period of revolution). In the case of the Moon, the periods of ω and Ω are respectively nine and eighteen years. The period of a long-period term is generally of the order of the period of revolution divided by a small parameter ε which characterizes the disturbing function.

Such a term is obtained by the integration (with respect to time) of a term for which $\alpha = \delta = 0$. Let us assume that this term is obtained in the second step of the calculation, i.e. that its coefficient is of the order of ε. We assume, more specifically, that this term appears as $\varepsilon A \cos(\beta\bar{\omega} + \gamma\bar{\Omega})$ in the equation in $d\Omega/dt$. Integration then gives

$$\frac{\varepsilon A \sin(\beta\bar{\omega} + \gamma\bar{\Omega})}{\beta n_\omega + \gamma n_\Omega} \tag{55}$$

Furthermore, n_ω and n_Ω themselves are of the order of ε, and therefore the corresponding term of Ω is of zero order in ε.

We can thus see that the order of long-period terms is always smaller by one unit than that of the short-period terms obtained in the same approximation. Thus, if the solution is to be calculated up to ε^2, we must proceed with the calculation of the long-period terms up to ε^3 (the coefficients of these are comparable with those of most second-order terms).

53. Convergence of the Series of the Solution

We have seen in section 44 that the series in the expansion of the disturbing function converge when ε is sufficiently small. The same applies to the various partial differential derivatives of R, and therefore to the right-hand sides of the differential equations involving the osculating elements. The question now is whether the same also applies to the series of the solution described above.

This problem has been investigated by several authors, particularly by Poincaré. This work can be summed up qualitatively as follows:

From the very first step of this solution (section 51.b), each term of the integrated series (49) and (50) has involved the divisor

$$D = \sum_j \alpha_j n_{j_0} \quad j > 1$$

where n_{j_0} (mean motion) is any number depending only on the initial conditions of the system, and the α_j terms are positive or negative integers, or zero, but not all can be zero.

In an infinite number of cases, D is as close to zero as we wish. Thus, if there are only two angular variables, $D = \alpha_1 n_1 + \alpha_2$ (n_2 being taken as 1), $-\alpha_1/\alpha_2$ can take an infinite number of values equal to any rational number. We know further that an infinite number of rational numbers exists between $n_2/n_1 - \varepsilon$ and $n_2/n_1 + \varepsilon$, whatever the value of ε.

It follows that a necessary condition of convergence is that the coefficients b or B must be sufficiently small to compensate for the smallness of the divisors. One can show that there is a number n', as close to n_1 as we wish, such that $B_{\alpha_1\alpha_2}/(\alpha_1 n_1 + \alpha_2)$ is not bounded, whatever the value of $B_{\alpha_1\alpha_2}$ (not zero). Since the value of n given by observation is not infinitely accurate, we can always find a number n' which gives a good representation of the observations, and for which there is at least one term in the series in question that is not bounded.

However, we can choose in the same interval another n'' for which the series uniformly converges. In fact, it follows from d'Alembert's property that the coefficient $B_{\alpha_1\alpha_2}$ is such that

$$|B_{\alpha_1\alpha_2}| < Ke^{\alpha_1}e'^{\alpha_2}$$

where e and e' are less than one, and K is a finite number. We now put $n'' = \sqrt{p/q}$, where p and q are two relative prime integers, such that pq is not a perfect square. We can always choose p and q so that n'' is made as close to n as we wish. We then have that

$$\left| \frac{1}{\alpha_1 n - \alpha_2} \right| = \left| \frac{\alpha_1 n + \alpha_2}{\alpha_1^2 n^2 - \alpha_2^2} \right| = \left| \frac{(\alpha_1 n + \alpha_2)q}{p\alpha_1^2 - q\alpha_2^2} \right| < q(|\alpha_1|n + |\alpha_2|)$$

since the denominator is a non zero integer. Hence

$$\left|\frac{B_{\alpha_1\alpha_2}}{D}\right| < Kq\left(|\alpha_1|\,n + |\alpha_2|\right)e^{\alpha_1}e'^{\alpha_2}$$

The series of the general term $\alpha_1 e^{\alpha_1} e'^{\alpha_2}$ or $\alpha_2 e^{\alpha_1} e'^{\alpha_2}$ derived from the series in the general term $e^{\alpha_1} e'^{\alpha_2}$ converge, and hence the series in question converges absolutely and uniformly.

It can therefore always be ensured that the series (49) and (50) converge, remaining within the limits of accuracy of the observations.

The above results show that these series cannot converge for integration constants situated in a continuous set of values, since the interval will include values of n equal to n' or n''.

The situation is similar as regards the convergence of the series (51) and (52) after the approximations leading to the final formal solution: the convergence of each approximation is a function of the convergence of the previous approximation.

If we put as a further condition that the series should converge for a whole range of the small parameter ε of the disturbing function, it has been shown that the series generally diverge instead. However, the problem has not been resolved in all the cases.

NOTE:

There is no great practical importance attached to the problem of convergence of the complete formal series. It is sufficient to calculate with the numerical values of the integration constants and the parameters a finite expression differing from the solution only by a quantity η as small (as we wish to make it) in a finite time interval Δt. And it has been shown by Poincaré that this is possible.

ʼ

CHAPTER V

THE MOTION OF AN ARTIFICIAL SATELLITE

The preceding Chapter gave a general method of expanding the disturbing function and the construction of a formal solution which, according to the Delaunay-Tisserand theorem, is in the form of a Fourier series with several linear arguments with respect to time and constant coefficients, plus possibly a linear function of time. It was remarked that while these series are generally divergent and truncated, they nevertheless allow a representation of the solution in a finite time interval Δt.

In the present Chapter we shall construct such a solution on a simple example, using a method, described by Von Zeipel, which leads rapidly to the solution. It should be noted, however, that any other method could also be used. In particular, it will be shown below that the method proposed in the last Chapter gives identical results.

Finally, encountering in a particular case the problem of small divisors which would make the formal series divergent, we shall see how this difficulty can be overcome.

Without attempting a complete solution of the proposed problem, we shall try here to give an example of the theory of celestial mechanics and to investigate some frequently encountered difficulties. The presented method could evidently be applied to more complicated cases, involving the motion of three or more bodies, but it is far from being the only one, and each particular case in celestial mechanics constitutes a specific problem.

54. The Potential of a Rigid Body

Consider a rigid body of finite volume V, in which at any point having coordinates ξ, η, ζ the density is $K(\xi, \eta, \zeta)$. Let there also be an external point P, of unit mass and coordinates x, y, z.

Each element Q of the solid attracts P according to Newton's law with a force

$$d\mathbf{F} = - k K (\xi, \eta, \zeta) \, d\xi \, d\eta \, d\zeta \frac{\mathbf{QP}}{QP^3}$$

whose three components are the partial derivatives of potential:

$$dU = k \frac{K (\xi, \eta, \zeta) \, d\xi \, d\eta \, d\zeta}{\sqrt{(x - \xi)^2 + (y - \eta)^2 + (z - \zeta)^2}}$$

(see section 7).

The assembly of elements Q attract P with a force

$$\mathbf{F} = \iiint_{(V)} d\mathbf{F}$$

whose three components are again partial derivatives of potential:

$$U = k \iiint_{(V)} \frac{K(\xi, \eta, \zeta) \, d\xi \, d\eta \, d\zeta}{\sqrt{(x - \xi)^2 + (y - \eta)^2 + (z - \zeta)^2}} \qquad (1)$$

The integration is carried out over the volume (V) of the rigid body. The potential U is thus a function only of x, y, and z.

The equations describing the motion of the point $P(x, y, z)$ subjected to this potential are

$$\frac{d^2x}{dt^2} = \frac{\partial U}{\partial x}; \quad \frac{d^2y}{dt^2} = \frac{\partial U}{\partial y}; \quad \frac{d^2z}{dt^2} = \frac{\partial U}{\partial z} \qquad (2)$$

NOTE:

These are the same as formulae (11) of Chapter I, except that here we are not dealing with n discrete bodies but with an assembly of elements constituting a single body.

55. Expansion of the Potential

Let r be the distance of P from the coordinate origin O $(r^2 = x^2 + y^2 + z^2)$, ρ the distance of a particle Q of the body from $O(\rho^2 = \xi^2 + \eta^2 + \zeta^2)$, S the angle (OP, OQ) and $\varDelta$ the distance PQ.

We have

$$\varDelta^2 = (x - \xi)^2 + (y - \eta)^2 + (z - \zeta)^2 = r^2 - 2r\rho \cos S + \rho^2$$

and, from section 45,

$$\frac{1}{\varDelta} = \frac{1}{r}\left[1 + \frac{\rho}{r}P_1(\cos S) + \frac{\rho^2}{r^2}P_2(\cos S) + \cdots + \frac{\rho^n}{r^n}P_n(\cos S) + \cdots\right]$$

where $P_n(\cos S)$ is a Legendre polynomial of order n.

Thus

$$U = \frac{k}{r}\iiint_{(V)} K(\xi, \eta, \zeta) \, d\xi \, d\eta \, d\zeta \left[1 + \sum_{j=1}^{\infty}\left(\frac{\rho}{r}\right)^j P_j(\cos S)\right]$$

which we shall write

$$U_0 + U_1 + U_2 + U_3 + \cdots + U_j + \cdots$$

(a) We have

$$U_0 = \iiint\limits_{(V)} K(\xi, \eta, \zeta)\, d\xi\, d\eta\, d\zeta = \frac{k}{r}\int\limits_V dM$$

but the above integral of the elements of mass taken over the whole volume of the body is equal to the mass of the body M, so that

$$U_0 = \frac{kM}{r} \tag{3}$$

(b) $U_1 = \int\limits_V \rho/r \cos S\, dM$, using the same simplified notation.

Now

$$\cos S = \frac{x\xi + y\eta + z\zeta}{\rho r}$$

whence

$$U_1 = \frac{k}{r^2}\int\limits_V (x\xi + y\eta + z\zeta)\, dM = \frac{k}{r^2}\left[x\int\limits_V \xi\, dM + y\int\limits_V \eta\, dM + z\int\limits_V \zeta\, dM \right]$$

If the origin lies at the centre of gravity of the body, each of the three integrals is by definition zero. We shall consequently assume that this is the case. Then

$$U_1 = 0 \tag{4}$$

(c)

$$U_2 = \frac{k}{r}\int\limits_V \frac{\rho^2}{r^2}\left[\tfrac{3}{2}\cos^2 S - \tfrac{1}{2}\right] dM$$

$$= \frac{k}{r^3}\int\limits_V \frac{1}{2}\left[3\frac{(x\xi + y\eta + z\zeta)^2}{r^2} - (\xi^2 + \eta^2 + \zeta^2) \right] dM$$

$$= \frac{k}{r^3}\left[\left(\frac{3}{2}\frac{x^2}{r^2} - \frac{1}{2}\right)\int\limits_V \xi^2\, dM + \left(\frac{3}{2}\frac{y^2}{r^2} - \frac{1}{2}\right)\int\limits_V \eta^2\, dM + \left(\frac{3}{2}\frac{z^2}{r^2} - \frac{1}{2}\right)\int\limits_V \zeta^2\, dM \right]$$

$$+ \frac{3k}{r^5}\left[xy\int\limits_V \xi\eta\, dM + yz\int\limits_V \eta\zeta\, dM + zx\int\limits_V \zeta\xi\, dM \right]$$

It will be recalled that, by definition, A, B, C are moments of inertia:

$$A = \int\limits_V (\eta^2 + \zeta^2)\, dM, \text{ etc...}$$

D, E, and F are products of inertia

$$D = \int_V \eta\zeta \, dM, \text{ etc...}$$

If we suppose that the coordinate axes are the principal inertial axes of the body

$$D = E = F = 0$$

Moreover

$$A + B + C = 2\int (\xi^2 + \eta^2 + \zeta^2) \, dM$$

whence

$$\int \xi^2 \, dM = \frac{A + B + C}{2} - A$$

We then have

$$U_2 = \frac{k}{r^3}\left[\left(\frac{3}{2}\frac{x^2}{r^2} - \frac{1}{2}\right)\left(\frac{A + B + C}{2} - A\right)\right.$$
$$+ \left(\frac{3}{2}\frac{y^2}{r^2} - \frac{1}{2}\right)\left(\frac{A + B + C}{2} - B\right) + \left.\left(\frac{3}{2}\frac{z^2}{r^2} - \frac{1}{2}\right)\left(\frac{A + B + C}{2} - C\right)\right]$$
$$U_2 = \frac{k}{r^3}\left[\frac{1}{2}(A + B + C) - \frac{3}{2}\left(\frac{Ax^2 + By^2 + Cz^2}{r^2}\right)\right]$$

Changing over to polar coordinates (longitude θ and latitude φ):

$$x = r\cos\theta\cos\varphi; \quad y = r\sin\theta\cos\varphi; \quad z = r\sin\varphi.$$

we obtain after some simple calculations

$$U_2 = \frac{k}{r^3}\left[\left(C - \frac{A + B}{2}\right)(\tfrac{1}{2} - \tfrac{3}{2}\sin^2\varphi) - \tfrac{3}{4}(A - B)\cos^2\varphi\cos 2\theta\right] \qquad (5)$$

If we further postulate that the body has a rotational symmetry about the $O\zeta$ axis, then $A = B$ and

$$U_2 = \frac{k}{r^3}(C - A)(\tfrac{1}{2} - \tfrac{3}{2}\sin^2\varphi)$$

Note that we again encounter a Legendre polynominal $P_2(\sin\varphi)$.

The quantity $\tfrac{3}{2}(C - A)/Ma_e^2 = J$, applied to the Earth, where a_e is the equatorial radius, M the mass of the Earth, and A, C its principal moments of inertia, is a fundamental geophysical and geodesic quantity. It is connected with the flattening of the Earth, ε.

We put $J_2 = \tfrac{2}{3}Ja_e^2$*, and

$$U_2 = \frac{kM}{r^3}J_2(\tfrac{1}{2} - \tfrac{3}{2}\sin^2\varphi)$$

* Since this book was written, the general trend was to set $J_2 = -\tfrac{2}{3}J$. Therefore, the formulae given in this chapter are consistent with the usual definition if it is assumed that $a_e = 1$.

(d) The calculation of subsequent quantities is carried out in an analogous manner. Nevertheless, if we suppose that the solid has an equatorial plane of symmetry, the potentials of odd order are zero and it can be shown that the potential of a solid of revolution is

$$U = \frac{kM}{r}\left[1 + \left(\frac{1}{r}\right)^2 J_2\left(\tfrac{1}{2} - \tfrac{3}{2}\sin^2\varphi\right) + \left(\frac{1}{r}\right)^4 J_4\left(\tfrac{3}{8} - \tfrac{15}{4}\sin^2\varphi + \tfrac{35}{8}\sin^4\varphi\right) + \cdots\right] \quad (6)$$

having the general form

$$U = \frac{kM}{r}\left[1 - \sum_{n=1}^{\infty} \frac{1}{r^{2n}} J_{2n} P_{2n}(\sin\varphi)\right] \quad (7)$$

in which J_{2n} is a numerical coefficient of P_{2n}, a Legendre polynomial of order $2n$.

56. The Case of a Nearly Spherical Body

A body having a spherical symmetry is distinguished by the fact that its three principal moments of inertia are equal and that its three principal inertial axes are any three mutually perpendicular axes of the body. Thus $J_2 = 0$. It could also be shown that all the J_{2n} are zero. This important fact can however be rapidly demonstrated.

Such a body can be regarded as made up of an infinite number of solid spherical layers of radius R, homogeneous, infinitely thin, and centred on O. If σ is the surface density of such a layer, its mass will be $m = 4\pi R^2\sigma$.

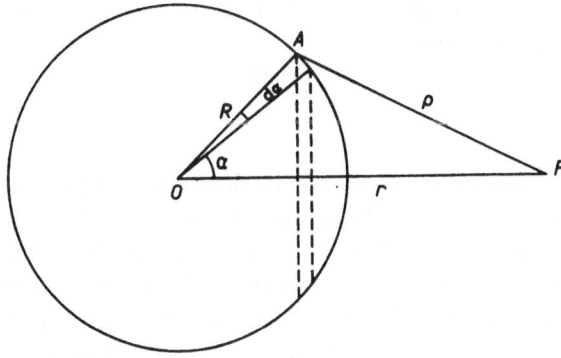

Fig. 8.

Let us calculate the potential of this layer at a point P. Let r be the distance from O to P (Figure 8), and consider the elementary zone between the circles defined by angles α and $\alpha + d\alpha$. The surface of this zone is:

$$dS = 2\pi R^2 \sin\alpha \, d\alpha$$

and its potential at P is

$$dU = \frac{k\sigma \, dS}{\rho}$$

In the triangle OAP:

$$\rho^2 = r^2 + R^2 - 2Rr \cos \alpha$$

whence

$$dU = \frac{2k\pi R^2 \sigma \sin \alpha \, d\alpha}{\sqrt{r^2 + R^2 - 2Rr \cos \alpha}}$$

$$U = 2k\pi R^2 \sigma \int_0^{\pi} \frac{\sin \alpha \, d\alpha}{\sqrt{r^2 + R^2 - 2Rr \cos \alpha}}$$

The quantity under the integral sign is an exact differential, and

$$U = \frac{2k\pi R\sigma}{r} \left[\sqrt{r^2 + R^2 - 2Rr \cos \alpha} \right]_0^{\pi}$$

If the point P lies *outside* the sphere, the expression in the square brackets has extreme values $r + R$ and $r - R$, so that

$$U = \frac{4k\pi R^2 \sigma}{r} = \frac{km}{r}$$

The potential – and thus also the forces – is the same as if all the mass were concentrated at O. This gives us the important result, obtained by summing up the effects of all the spherical homogeneous layers making up the body having a spherical symmetry: a body of this kind behaves as if all its mass were concentrated at its centre. Its potential is therefore:

$$U = \frac{kM}{r} \tag{8}$$

NOTE:

The above proof also makes it possible to show that the potential of a spherical shell is constant inside the shell. This result will not be used further in this book.

Formula (8) is inexact for the case of a nearly spherical body, but the deviations are small. Thus, the quantities $J_2, J_4, ..., J_{2n}$ of formula (7) are small. If r increases, the effect of the correction terms falls off as $1/r^{2n}$. For this reason in the study of the motion of nearly spherical bodies such as the planets we can neglect these correction terms and assume that these bodies behave as if all their mass were concentrated at their centre of gravity (section 5).

57. Equations of Motion of an Artificial Satellite

In contrast to the case of natural bodies, whose separations are very large in comparison to their size and to which the results of the preceding section may be applied, the artificial Earth satellites are so close that the secondary terms of the Earth's potential can no longer be neglected. The same situation is encountered with some natural satellites – Phobos, Deimos, Jupiter V, etc. – which are subject to considerable perturbations of this kind. However, the imprecisions in the observations of the motion of natural bodies were such that very approximate theory was sufficient to account for the observed deviations. In contrast, the accuracy of the observations of the terrestrial artificial satellites is several orders of magnitude higher, and for this reason the study of the motion of satellites in a potential field of a nearly spherical body developed particularly strongly since the launching of the first sputniks.

Thus, if the forces exerted on a satellite are of gravitational origin, depending only on the potential U (7) of the planet, the equations of motion are those given by (2):

$$\frac{d^2x}{dt^2} = \frac{\partial U}{\partial x}; \quad \frac{d^2y}{dt^2} = \frac{\partial U}{\partial y}; \quad \frac{d^2z}{dt^2} = \frac{\partial U}{\partial z} \tag{9}$$

The results of Chapter III are of course applicable, and the preceding system is equivalent to a Delaunay system:

$$\frac{dL}{dt} = \frac{\partial \Phi}{\partial l}; \quad \frac{dG}{dt} = \frac{\partial \Phi}{\partial g}; \quad \frac{dH}{dt} = \frac{\partial \Phi}{\partial h}$$

$$\frac{dl}{dt} = -\frac{\partial \Phi}{\partial L}; \quad \frac{dg}{dt} = -\frac{\partial \Phi}{\partial G}; \quad \frac{dh}{dt} = -\frac{\partial \Phi}{\partial H} \tag{10}$$

with

$$\Phi = \frac{\mu}{2a} + R$$

and

$$R = U - \frac{kM}{r}$$

or to a system of Lagrange equations [system (45) of section 32].

However, apart from experiencing terrestrial gravitational forces, an artificial satellite is perturbed by the Moon and the Sun, and is braked by friction with the upper atmosphere, the latter effect increasing with decreasing altitude of the satellite. This force is very difficult to analyze, as the atmospheric density depends on temperature, illumination, solar activity, and so on. An artificial satellite is also subject to pressure exerted by solar radiation, and this effect may sometimes be very significant because it is proportional to the ratio S/M of the illuminated surface to the mass of the satellite. Although radiation pressure is also exerted on other bodies in the solar system, it is negligible because in these cases the ratio S/M is extremely small. Finally, the trajectory of an artificial satellite may be perturbed in more or less predictable fashion by electromagnetic effects, impact of meteorites, etc.

Below we shall be concerned only with the effects of the terrestrial gravitational potential, the primary cause of the perturbations of a high satellite orbit. Detailed study of this motion will allow us to describe methods applicable to most other problems encountered in celestial mechanics. The main differences characterizing these principal problems will be indicated at the appropriate points.

58. The Principle of Von Zeipel's Method

The equations

$$
\left.
\begin{aligned}
\frac{dL}{dt} &= \frac{\partial \Phi}{\partial l}; & \frac{dG}{dt} &= \frac{\partial \Phi}{\partial g}; & \frac{dH}{dt} &= \frac{\partial \Phi}{\partial h} \\
\frac{dl}{dt} &= -\frac{\partial \Phi}{\partial L}; & \frac{dg}{dt} &= -\frac{\partial \Phi}{\partial G}; & \frac{dh}{dt} &= -\frac{\partial \Phi}{\partial H}
\end{aligned}
\right\}
\tag{11}
$$

where Φ is a function of six Delaunay variables, will be solved by a series of variable transformations with the aid of a determining function. Applying the results of section 23 to system (11), we see that if we consider a new system of variables L', G', H', l', g', h', and a determining function

$$
S(L', G', H', l, g, h)
$$

such that

$$
L = \frac{\partial S}{\partial l}; \quad G = \frac{\partial S}{\partial g}; \quad H = \frac{\partial S}{\partial h}; \quad l' = \frac{\partial S}{\partial L'}; \quad g' = \frac{\partial S}{\partial G'}; \quad h' = \frac{\partial S}{\partial H'}
\tag{12}
$$

then the system of the new equations is canonical and the new Hamiltonian Φ', expressed in new variables, is unchanged.

Thus

$$
\Phi(L, G, H, l, g, h) = \Phi'(L', G', H', l', g', h')
\tag{13}
$$

and

$$
\left.
\begin{aligned}
\frac{dL'}{d} &= \frac{\partial \Phi'}{\partial l'}; & \frac{dG'}{dt} &= \frac{\partial \Phi'}{\partial g'}; & \frac{dH'}{dt} &= \frac{\partial \Phi'}{\partial h'} \\
\frac{dl'}{dt} &= -\frac{\partial \Phi'}{\partial L'}; & \frac{dg'}{dt} &= -\frac{\partial \Phi'}{\partial G'}; & \frac{dh'}{dt} &= -\frac{\partial \Phi'}{\partial H'}
\end{aligned}
\right\}
\tag{14}
$$

Since S is any function, we can impose on it certain conditions. In the method which will be treated in this section, introduced by Von Zeipel and developed by Brouwer, the condition is that Φ' is such as to be independent of one of the angular variables on which Φ depends. If this condition can be satisfied, and we shall demonstrate that it can, we shall thus eliminate one angular variable from the Hamiltonian. If this operation is repeated several times, all the angular variables will be eliminated from Φ one by one, and the final Hamiltonian Φ'' will be a function of only L'', G'', and H''.

In the first three final equations

$$
\frac{dL''}{dt} = \frac{\partial \Phi''}{\partial l''}; \quad \frac{dG''}{dt} = \frac{\partial \Phi''}{\partial g''}; \quad \frac{dH''}{dt} = \frac{\partial \Phi''}{\partial h''}
$$

the right-hand sides will be zero, and consequently L'', G'' and H'' will be constants in the solution. Putting the values of these constants back into $\partial\Phi''/\partial L'$, $\partial\Phi''/\partial G''$, and $\partial\Phi''/\partial H''$, we find that dl''/dt, dg''/dt, and dh''/dt are constants. In the solution, l'', g'', and h'' will be linear functions of time.

We shall return to the initial variables L, G, H, l, g, and h with the aid of various equations of type (12) which define each intermediate variable.

59. Setting Up of the Equations

We shall now consider the operations which allow a solution of equations (9). To reduce the volume of algebraic calculations, we shall limit ourselves to the first term of the expansion of U, i.e.

$$R = U_2 = \frac{\mu}{r^3} J_2 (\tfrac{1}{3} - \tfrac{3}{2} \sin^2 \varphi) \quad (\mu = kM)$$

Addition of further terms does not make any essential difference to the method.

In the system given by (11), Φ is expressed as a function of Delaunay variables. We shall not complete this step; it could be accomplished by the methods described in the preceding Chapter on the expansion of the disturbing function.

We have

$$\Phi = \frac{\mu}{2a} + \frac{\mu}{r^3} J_2 (\tfrac{1}{3} - \tfrac{3}{2} \sin^2 \varphi)$$

In a spherical triangle defined by the plane of the orbit, the equational plane, and the plane of the meridian passing through the satellite (Figure 9):

$$\sin \varphi = \sin i \sin (g + v)$$

where v is the true anomaly.

Therefore

$$\sin^2 \varphi = \sin^2 i \, \frac{1 - \cos (2g + 2v)}{2}$$

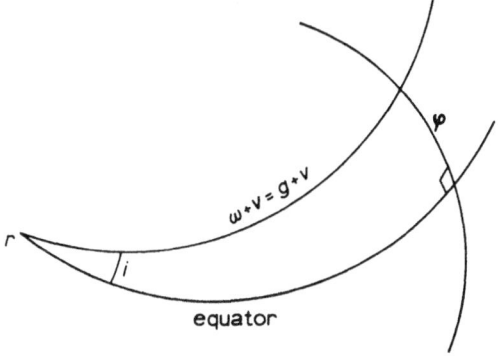

equator

Fig. 9.

Now, from formulae (42) of section 30,

$$\cos i = \frac{G}{H} \quad \text{and} \quad \sqrt{a} = \frac{L}{\sqrt{\mu}}$$

whence

$$\Phi = \frac{\mu^2}{2L^2} + \frac{\mu^4 J_2}{L^6}\left[\frac{a^3}{r^3}\left(-\frac{1}{4} + \frac{3}{4}\frac{H^2}{G^2}\right) + \left(\frac{3}{4} - \frac{3}{4}\frac{H^2}{G^2}\right)\frac{a^3}{r^3}\cos(2g + 2v)\right] \quad (15)$$

Note that now Φ does not contain h.

The methods available for the expansion of the functions relating to the two-body problem, which were already given in section 40, allow a^3/r^3 and $a^3/r^3 \cos(2g + 2v)$ to be expressed as a function of e, g, and l. Since $e = \sqrt{1 - (G^2/L^2)}$, Φ is thus expressed as a function of Delaunay variables. Neglecting e^4, assumed to be small:

$$\frac{a^3}{r^3} = 1 + \frac{3e^2}{2} + (3e + \tfrac{27}{8}e^3)\cos l + \tfrac{9}{2}e^2\cos 2l + \tfrac{53}{8}e^3\cos 3l$$

$$\frac{a^3}{r^3}\cos(2g + 2v) = \frac{e^3}{48}\cos(2g - l) + (-\frac{e}{2} + \tfrac{1}{16}e^3)\cos(2g + l)$$

$$+ (1 - \tfrac{5}{2}e^2)\cos(2g + 2l) + (\tfrac{7}{2}e - \tfrac{123}{16}e^3)\cos(2g + 3l) + \tfrac{17}{2}e^2\cos(2g + 4l)$$

$$+ \tfrac{845}{48}e^3\cos(2g + 5l)$$

There are two quantities of particular importance, and these will be calculated directly.

(a) The constant term of the expansion of a^3/r^3 is L^3/G^3. According to Fourier's theorem (section 35), this term is

$$A_0 = \frac{1}{2\pi}\int_0^{2\pi}\frac{a^3}{r^3}\,dM$$

Now from the law of areas,

$$r^2\frac{dv}{dt} = na^2\sqrt{1 - e^2} = G$$

and

$$\frac{dM}{dt} = n = \frac{\sqrt{\mu}}{a^{3/2}} = \frac{L}{a^2}$$

so that

$$\frac{r^2}{a^2}\frac{dv}{dM} = \frac{G}{L}$$

Moreover

$$A_0 = \frac{1}{2\pi}\int_0^{2\pi}\frac{a}{r}\times\frac{L}{G}\,dv = \frac{1}{2\pi}\frac{L^3}{G^3}\int_0^{2\pi}(1 + e\cos v)\,dv = \frac{L^3}{G^3} = (1 - e^2)^{-3/2}$$

(b) The term in cos $2g$ in the expansion of

$$\frac{a^3}{r^3} \cos(2g + 2v)$$

is zero. According to Fourier's theorem this term is

$$C_0 = \frac{1}{\pi} \int_0^{2\pi} \frac{a^3}{r^3} \cos(2g + 2v) \, dM = \frac{1}{\pi} \int_0^{2\pi} \frac{L}{G} \frac{a}{r} \cos(2g + 2v) \, dv$$

$$= \frac{1}{2\pi} \frac{L^3}{G^3} \int_0^{2\pi} (1 + e \cos v) \cos(2g + 2v) \, dv = 0$$

60. Elimination of the Mean Anomaly

Following the method outlined in section 58 we shall try to determine the determining function S and the new Hamiltonian Φ' which does not contain explicitly the new variable l'. We shall assume J_2 to be small, and following Poincaré's theorem (sections 48 and 49) shall seek a solution which can be expanded into a complete series in J_2. We shall also suppose that S and Φ' can be expanded in the same way. We put

$$\Phi = \Phi_0 + \Phi_1$$

The subscripts indicate the order of the terms in J_2, so that

$$\Phi_0 = \frac{\mu^2}{2L^2}$$

$$\Phi_1 = \frac{\mu^4 J_2}{L^6} \left[\left(-\frac{1}{4} + \frac{3}{4} \frac{H^2}{G^2} \right) \frac{a^3}{r^3} + \left(\frac{3}{4} - \frac{3}{4} \frac{H^2}{G^2} \right) \frac{a^3}{r^3} \cos(2g + 2v) \right]$$

In the same way

$$\left. \begin{aligned} \Phi' &= \Phi'_0 + \Phi_1 + \Phi'_2 + \cdots \\ S &= S_0 + S_1 + S_2 + \cdots \end{aligned} \right\} \tag{16}$$

One requirement is that, when J_2 is neglected, the change of the variables should lead to an identity. We therefore put $S_0 = L'l + G'g + H'h$, and have, in accordance with (12),

$$\left. \begin{aligned} L &= \frac{\partial S}{\partial l} = L' + \frac{\partial S_1}{\partial l} + \frac{\partial S_2}{\partial l} + \cdots; \quad l' = \frac{\partial S}{\partial L'} = l + \frac{\partial S_1}{\partial L'} + \frac{\partial S_2}{\partial L'} + \cdots \\ G &= \frac{\partial S}{\partial g} = G' + \frac{\partial S_1}{\partial g} + \frac{\partial S_2}{\partial g} + \cdots; \quad g' = \frac{\partial S}{\partial G'} = g + \frac{\partial S_1}{\partial G'} + \frac{\partial S_2}{\partial G'} + \cdots \\ H &= \frac{\partial S}{\partial h} = H' + \frac{\partial S_1}{\partial h} + \frac{\partial S_2}{\partial h} + \cdots; \quad h' = \frac{\partial S}{\partial H'} = h + \frac{\partial S_1}{\partial H'} + \frac{\partial S_2}{\partial H'} + \cdots \end{aligned} \right\} \tag{17}$$

Equating Φ with Φ' (eq. 13 in section 58) shows that

$$\Phi_0(L) + \Phi_1(L, G, H, l, g, -)$$
$$\equiv \Phi_0' + \Phi_1'(L', G', H', -, g', -) + \Phi_2'(L', G', H', - g' -) + \cdots$$

where dashes indicate missing variables and where Φ' no longer depends on l'. Since Φ_1 is already independent of h, we attempt to find a Φ' which depends only on g'. Replacing L, G, H, l', g', and h' by their values expressed as a function of S, we find that

$$\Phi_0\left(\frac{\partial S}{\partial l}\right) + \Phi_1\left(\frac{\partial S}{\partial l}, \frac{\partial S}{\partial g}, \frac{\partial S}{\partial g}, l, g, -\right)$$
$$\equiv \Phi_0' + \Phi_1'\left(L', G', H', -, \frac{\partial S}{\partial G'}, -\right) + \Phi_2'\left(L', G', H', -, \frac{\partial S}{\partial G'}, -\right) + \cdots$$

Replacement of the partial derivatives of S by their expansions in J_2 and truncation of the expansion, for example, at terms in J_2^2 leads to

$$\Phi_0\left(L' + \frac{\partial S_1}{\partial l} + \frac{\partial S_2}{\partial l}\right) + \Phi_1\left(L' + \frac{\partial S_1}{\partial l}, G' + \frac{\partial S_1}{\partial g}, H' + \frac{\partial S_1}{\partial h}, l, g, -\right)$$
$$\equiv \Phi_0' + \Phi_1'\left(L', G', H', -, g + \frac{\partial S_1}{\partial G'}, -\right) + \Phi_2'(L', G', H', -, g, -)$$

Expansion in Taylor series gives

$$\Phi_0(L') + \frac{\partial \Phi_0}{\partial L'}\frac{\partial S_1}{\partial l} + \frac{\partial \Phi_0}{\partial L'}\frac{\partial S_2}{\partial l} + \frac{1}{2}\frac{\partial^2 \Phi_0}{\partial L'^2}\left(\frac{\partial S_1}{\partial l}\right)^2 + \Phi_1(L', G', H', -, g, -)$$
$$+ \frac{\partial \Phi_1}{\partial L'}\frac{\partial S_1}{\partial l} + \frac{\partial \Phi_1}{\partial G'}\frac{\partial S_1}{\partial g} + \frac{\partial \Phi_1}{\partial H'}\frac{\partial S_1}{\partial h} \equiv \Phi_0' + \Phi_1'(L', G', H', -, g, -)$$
$$+ \frac{\partial \Phi_1}{\partial g}\frac{\partial S_1}{\partial G'} + \Phi_2'(L', G', H', -, g, -)$$

Collecting and equating terms of the same order of J_2, we see that

(a) $\qquad \Phi_0' = \Phi_0(L')$

(b) $\qquad \Phi_1'(L', G', H', -, g, -) = \frac{\partial \Phi_0}{\partial L'}\frac{\partial S_1}{\partial l} + \Phi_1(L', G', H', l, g, -)$

(c) $\qquad \Phi_2' + \frac{\partial \Phi_1}{\partial g}\frac{\partial S_1}{\partial G'} + \frac{\partial \Phi_0}{\partial L'}\frac{\partial S_2}{\partial l}$

$\qquad\qquad + \frac{1}{2}\frac{\partial^2 \Phi_0}{\partial L'^2}\left(\frac{\partial S_1}{\partial l}\right)^2 + \frac{\partial \Phi_1}{\partial L'}\frac{\partial S_1}{\partial l} + \frac{\partial \Phi_1}{\partial G'}\frac{\partial S_1}{\partial g} + \frac{\partial \Phi_1}{\partial H'}\frac{\partial S_1}{\partial h}$

(18)

The first equation (a) gives

$$\Phi_0' = \frac{\mu^2}{2L'^2}$$

The second equation (b) can be dealt with in two steps. We know that Φ_1 expanded in l has only one term that is independent of l, and it contains no term in $\cos 2g$. Separating the only term independent of l, (Φ_{1S}), we write that

$$\Phi_1 = \frac{\mu^4 J_2}{L'^6}\left(-\frac{1}{4} - \frac{3}{4}\frac{H'^2}{G'^2}\right)\frac{L'^3}{G'^3} + \Phi_{1p} = \Phi_{1S} + \Phi_{1p}$$

(The subscripts S and p denote secularity and periodicity.)

We shall now assume that S depends on l only through the trigonometric terms (form analogous to that of the solution). In this case, $\partial S_1/\partial l$ has no constant term. We can therefore split equation (b) in two, and thus obtain an equation involving terms in l:

$$\frac{\partial \Phi_0}{\partial L'}\frac{\partial S_1}{\partial l} + \Phi_{1p}(L', G', H', l, g, -) = 0$$

and another equation involving only secular terms:

$$\Phi_1'(L', G', H', -, g, -) = \Phi_{1S}(L', G', H', -, g, -) = \frac{\mu^4 J_2}{L'^3 G'^3}\left(-\frac{1}{4} + \frac{3}{4}\frac{H'^2}{G'^2}\right) \qquad (19)$$

The second equation defines Φ_1', while on account of ·

$$\frac{\partial \Phi_0}{\partial L'} = \frac{-\mu^2}{L'^3},$$

the first equation gives

$$\frac{\partial S_1}{\partial l} = -\frac{L'^3}{\mu^2}\Phi_{1p}(L', G', H', l, g, -)$$

and

$$S_1 = \int \frac{L'^3}{\mu^2}\Phi_{1p}(L', G', H', l, g, -)\,dl \qquad (20)$$

We have thus obtained S_1 (independent of h) and Φ_1' (independent of l' and h'). We can substitute them in (18c) and then, equating separately the terms containing l and those independent of it, find S_2 and Φ_2'; we could find higher S and Φ' terms if we had continued the expansion in J_2 leading to eq. (18). Finally, the Hamiltonian Φ_2' depends on g', but no more on l' and h'.

61. Explicit Expression for S_1

To make the change of variables (17), explicit, we must give an explicit expression for S, which we shall do by calculating only S_1 as defined by (20).

We have already seen that

$$\Phi_{1S} = \frac{\mu^4 J_2}{L'^6}\left(-\frac{1}{4} + \frac{3}{4}\frac{H'^2}{G'^2}\right)\frac{L'^3}{G'^3}$$

Whence, using (15):

$$\Phi_{1p}(L', G', H', l, g -)$$
$$= \frac{\mu^4 J_2}{L'^6}\left[\left(-\frac{1}{4}+\frac{3}{4}\frac{H'^2}{G'^2}\right)\left(\frac{a'^3}{r'^3}-\frac{L'^3}{G'^3}\right)+\left(\frac{3}{4}-\frac{3}{4}\frac{H'^2}{G'^2}\right)\frac{a'^3}{r'^3}\cos(2g+2v')\right]$$

where a', r', and v' are known functions of the elliptical motion expressed as a function of the metric elements L' and G' and the angular element l. Knowing that

$$dl = n'dt = \frac{r'^2}{a'^2\sqrt{1-e'^2}}\,dv' = \frac{L'}{G'}\frac{r'^2}{a'^2}\,dv'$$

we must calculate

$$\int\left(\frac{a'^3}{r'^3}-\frac{L'^3}{G'^3}\right)dl = \frac{-L'^3}{G'^3}\,l+\frac{L'^3}{G'^3}\int(1+e'\cos v')\,dv'$$
$$= \frac{L'^3}{G'^3}(v'-l+e'\sin v')$$

We deal similarly with the rest of the expression, and hence

$$S_1 = \frac{\mu^2 J_2}{G'^3}\left[\left(-\frac{1}{4}+\frac{3}{4}\frac{H'^2}{G'^2}\right)(v'-l+e'\sin v')+\left(\frac{3}{4}-\frac{3}{4}\frac{H'^2}{G'^2}\right)\right.$$
$$\left.\left(\tfrac{1}{2}\sin(2g+2v')+\frac{e'}{2}\sin(2g+v')+\frac{e'}{6}\sin(2g+3v')\right)\right] \quad (21)$$

where v' is the true anomaly on the ellipse with elements L', G' and l. The change of variables in (17) is thus defined to within J_2; calculations involving also J_2^2 are naturally more complicated. We also note that $\partial S_1/\partial h = 0$ and thus $H' = H$.

62. Calculation of Φ'_2

Knowing S_1, we can separate in (18c) the terms independent of l. We now expand (21) as a function of l. However, by virtue of Fourier's theorem (cf. section 59), we can also perform the calculations directly, and thus find the terms independent of l. Nevertheless, this calculation is long, and we shall not reproduce it here.
We find that

$$\Phi'_2 = \frac{\mu^6 J_2^2}{L'^{10}}\left[\frac{15}{128}\frac{L'^5}{G'^5}\left(1-\frac{18}{5}\frac{H'^2}{G'^2}+\frac{H'^4}{G'^4}\right)\right.$$
$$+\frac{3}{32}\frac{L'^6}{G'^6}\left(1-6\frac{H'^2}{G'^2}+9\frac{H'^4}{G'^4}\right)-\frac{15}{128}\frac{L'^7}{G'^7}\left(1-2\frac{H'^2}{G'^2}+9\frac{H'^4}{G'^4}\right)$$
$$\left.-\frac{3}{64}\left(\frac{L'^5}{G'^5}-\frac{L'^7}{G'^7}\right)\left(1-16\frac{H'^2}{G'^2}+15\frac{H'^4}{G'^4}\right)\cos 2g'\right] = \Phi'_{2S}+\Phi'_{2p} \quad (22)$$

where Φ'_{2p} contains the $\cos 2g'$ term, and Φ'_{2S} involves the term that is independent of g'.

63. Elimination of g

We have thus reduced the set of equations to the canonical set (14) whose Hamiltonian is given by

$$\Phi' = \Phi'_0 + \Phi'_1 + \Phi'_{2s} + \Phi'_{2p}$$

where only Φ'_{2p} depends on g', the others depending only on L', G', and H'.

We shall now use the determining function S' to perform another change of canonical variables:

$$L', G', H', l', g', h' \rightarrow L'', G'', H'', l'', g'', h''$$

and have also that

$$S' = L''l' + G''g' + H''h' + S'_1(L'', G'', H'', g') + \cdots$$

As in the first approximation for h, it can be shown that S' is independent of l.

$$\left.\begin{array}{l} L = \dfrac{\partial S'}{\partial l'} = L''; \quad G' = \dfrac{\partial S'}{\partial g'} = G'' + \dfrac{\partial S'_1}{\partial g'} + \cdots; \quad H' = \dfrac{\partial S'}{\partial h'} = H'' \\[3mm] l'' = \dfrac{\partial S'}{\partial L''} = l' + \dfrac{\partial S'_1}{\partial L''} + \cdots; \quad g'' = \dfrac{\partial S'}{\partial G''} = g' + \dfrac{\partial S'_1}{\partial G''} + \cdots; \\[3mm] h'' = \dfrac{\partial S'}{\partial H''} = h' + \dfrac{\partial S'_1}{\partial H''} + \cdots \end{array}\right\} \qquad (23)$$

The new Hamiltonian $\Phi'' = \Phi''_0 + \Phi''_1 + \Phi''_2 + \ldots$ is independent of l'', g'', and h''. Its identification with Φ' leads to

$$\Phi''_0 + \Phi''_1 + \Phi''_2 = \Phi'_0 + \Phi'_1\left(L'', G'' + \frac{\partial S'_1}{\partial g'}, H''\right) + \Phi'_{2s} + \Phi'_{2p} \qquad (24)$$

where we stop at terms in J_2^2. This gives the equations:

$$\left.\begin{array}{l} \Phi''_0 = \Phi'_0(L'') \\[2mm] \Phi''_1 = \Phi'_1(L'', G'', H'') \\[2mm] \Phi''_2 = \dfrac{\partial \Phi'_1}{\partial G''}\dfrac{\partial S'_1}{\partial g'} + \Phi'_{2s} + \Phi'_{2p} \end{array}\right\}$$

the last of which reduces to two equations: terms independent of g'

$$\Phi''_2 = \Phi'_{2s}$$

and terms which depend on g':

$$\frac{\partial \Phi'_1}{\partial G''}\frac{\partial S'_1}{\partial g'} + \Phi'_{2p} = 0$$

This equation defines S_1:

$$S_1 = \int \frac{-\Phi'_{2p}(L'', G'', H'', -, g', -)}{\partial \Phi'_1 / \partial G''} \, dg'$$

Using (19) and (22), we find that

$$S_1' = \frac{\mu^2 J_2}{32 L''^2 G''} \frac{\left(1 - \dfrac{L''^2}{G''^2}\right)\left(1 - 16\dfrac{H''^2}{G''^2} + 15\dfrac{H''^4}{G''^4}\right)}{\left(1 - 5\dfrac{H''^2}{G''^2}\right)} \sin 2g' \qquad (25)$$

We obtain the following set of canonical equations with the aid of the new variables defined by this determining function S' and equations (23):

$$\frac{dL''}{dt} = \frac{\partial \Phi''}{\partial l''}; \qquad \frac{dG''}{dt} = \frac{\partial \Phi''}{\partial g''}; \qquad \frac{dH''}{dt} = \frac{\partial \Phi''}{\partial h''}$$

$$\frac{dl''}{dt} = -\frac{\partial \Phi''}{\partial L''}; \qquad \frac{dg''}{dt} = -\frac{\partial \Phi''}{\partial G''}; \qquad \frac{dh''}{dt} = -\frac{\partial \Phi''}{\partial H''}$$

where $\Phi'' = \Phi_0'' + \Phi_1'' + \Phi_2'' + \ldots$ now depends only on L'', G'', and H''. Moreover, $\partial \Phi''/\partial l''$, $\partial \Phi''/\partial g''$, and $\partial \Phi''/\partial h''$ are zero, and therefore, L'', G'', and H'' are constant.

The right-hand sides of the last three equations are also constant. On integration, l'', g'', and h'' are linear functions of time. The complete solution of the problem is thus in the form:

$$L'' = L_0; \qquad\qquad G'' = G_0; \qquad\qquad H'' = H_0$$
$$l'' = n_l(t - t_0); \quad g'' = n_g(t - t_0); \quad h'' = n_h(t - t_0)$$

Formulae (17) and (23) enable us to revert to the original variables L, G, H, l, g, and h, and to express any of these as a function of time.

64. Main Results: the Motion of Artificial Satellites

Neglecting J_2^2, we obtain simple expressions for l'', g'', and h''; it is sufficient to retain in Φ'' the following terms:

$$\Phi_0'' = \Phi_0' = \Phi_0 = \frac{\mu^2}{2L''^2}$$

and

$$\Phi_1'' = \Phi_1' = \Phi_{1S} = \frac{\mu^4 J_2}{L''^3 G''^3}\left(-\frac{1}{4} + \frac{3}{4}\frac{H''^2}{G''^2}\right)$$

$$\frac{dl''}{dt} = -\frac{\partial \Phi_0}{\partial L''} - \frac{\partial \Phi_1''}{\partial L''} = \frac{\mu^2}{L''^3} + \frac{3\mu^4 J_2}{L''^4 G''^3}\left(-\frac{1}{4} + \frac{3}{4}\frac{H''^2}{G''^2}\right)$$

$$\frac{dl''}{dt} = n_0 + 3n_0 \frac{J_2}{a_0^2}\left(-\frac{1}{4} + \frac{3}{4}\cos^2 i_0\right)(1 - e_0^2)^{-3/2} \qquad (26)$$

If $n_0 = \mu^2/L_0^3$ and e_0, a_0, and i_0 are the values of the elliptic elements corresponding to the constants L_0, G_0 and H_0

$$\frac{dg''}{dt} = -\frac{\partial \Phi_1''}{\partial G''} = \frac{-3\mu^4 J_2}{L''^3 G''^4}\left(\frac{1}{4} - \frac{5}{4}\frac{H''^2}{G''^2}\right) = n_0 \frac{J_2}{a_0^2}\left(-\frac{3}{4} + \frac{15}{4}\cos^2 i_0\right)\frac{1}{(1 - e_0^2)^2} \qquad (27)$$

$$\frac{dh''}{dt} = -\frac{\partial \Phi_1''}{\partial H''} = \frac{-3\mu^4 J_2 H''}{2L''^3 G''^5} = -\frac{n_0}{2}\frac{J_2}{a_0^2}\cos i_0 \times \frac{3}{(1 - e_0^2)^2} \qquad (28)$$

Reversion to the initial elements gives expressions of the following form:

$$l = l'' + l_L + l_c; \quad g = g'' + g_L + g_c; \quad h = h'' + h_L + h_c$$
$$L = L'' + L_L + L_c; \quad G = G'' + G_L + G_c; \quad H = H'' + H_L + H_c$$

where the first terms represent the solution described above, the terms with subscript L are the long-period periodic terms, g'' is involved only in the arguments of the trigonometric lines, the terms with subscript c are the short-period terms, and l'' is involved, alone or with g'', in the arguments of the trigonometric lines. Expressions (26) and (28) thus describe the secular part of the motion.

Eq. (28) shows that the node undergoes a retrograde displacement with a period of

$$\frac{4\,\pi\,(1 - e_0^2)^2\,a_0^2}{3\quad n_0 J_2\,\cos i_0}.$$

The movement is fastest when the inclination tends to zero, and is non-existent with polar orbits ($\cos i_0 = 0$). The motion tends to be slower with small eccentricities and large semi-major axes a_0.

The perigee describes a direct motion with satellites whose inclination is smaller than I_0, such that $5\cos^2 I_0 - 1 = 0$ ($I_0 = 63°26'$), its motion being retrograde when the inclination is greater.

The period of the satellite is close to $2\pi/n_0$, i.e. the value it would have if the Earth were a perfect sphere. The mean motion is more rapid with inclinations smaller than I_0', such that $3\cos^2 I_0' - 1 = 0$ ($I_0' = 54°44'$) and slower with larger inclinations.

In agreement with what was said in section 51, the semi-major axis, the eccentricity, and the inclination do not describe a secular motion. The terms in $\sin 2g''$ or $\cos 2g''$ coming from Φ_{2S}' (cf. section 62) are the long-period terms which may appear in all the elements, but are of the order of J_2 for all the elements except the semi-major axis. These are the terms whose period is half the period of the perigee. There are also terms of higher order in J_2: their periods are equal to a quarter, a sixth etc. of the period of the perigee.

The short-period terms distort the basic orbit which rotates and suffers long-period perturbations.

NOTE:

We have thus found again the distinction between long-period and short-period secular terms which was introduced in section 52. This distinction also exists in the theory of the motion of most natural bodies in the solar system.

65. Application of the Lagrange Equations; First Approximation

To apply the general method given in Section 51, we shall now attempt to find the perturbations of the elements of an artificial satellite using the Lagrange equations, but the search will be restricted to perturbations appearing in the first approximation.

The disturbing function R (eq. 15) is written in terms of the elliptical elements:

$$R = \frac{\mu J_2}{a^3}\left[(-\tfrac{1}{4} + \tfrac{3}{4}\cos^2 i)\frac{a^3}{r^3} + (\tfrac{3}{4} - \tfrac{3}{4}\cos^2 i)\frac{a^3}{r^3}\cos(2g + 2v)\right]$$

Using the expansions given in section 59 with the notations Ω, ω, and M for the elements, and neglecting e^3, we have that

$$R = \frac{\mu J_2}{a^3}\left\{(-\tfrac{1}{4} + \tfrac{3}{4}\cos^2 i)\left(1 + \frac{3e^2}{2} + 3e\cos M + \tfrac{9}{2}e^2\cos 2\dot M\right)\right.$$

$$+ (\tfrac{3}{4} - \tfrac{3}{4}\cos^2 i)\left[-\frac{e}{2}\cos(2\omega + M) + (1 - \tfrac{5}{2}e^2)\cos(2\omega + 2M)\right.$$

$$\left.\left. + \tfrac{7}{2}e\cos(2\omega + 3M) + \tfrac{17}{2}e^2\cos(2\omega + 4M)\right]\right\} \qquad (29)$$

To a first approximation, the elements a, e, i, Ω, and ω are constant in Keplerian motion, and $M = n(t - t_0)$ with $n^2 a^3 = \mu$. We shall now apply the Lagrange equation describing the movement of the perigee [eq. (45), section 34]:

$$\frac{d\omega}{dt} = \frac{\sqrt{1 - e^2}}{na^2 e}\frac{\partial R}{\partial e} - \frac{\cos i}{na^2\sqrt{1 - e^2}\sin i}\frac{\partial R}{\partial i}$$

and find that

$$\frac{\partial R}{\partial e} = \frac{\mu J_2}{a^3}\{(-\tfrac{1}{4} + \tfrac{3}{4}\cos^2 i)(3e + 3\cos M + 9e\cos 2M) + (\tfrac{3}{4} - \tfrac{3}{4}\cos^2 i)$$

$$\times [-\tfrac{1}{2}\cos(2\omega + M) - 5e\cos(2\omega + 2M) + \tfrac{7}{2}\cos(2\omega + 3M)$$

$$+ 17e\cos(2\omega + 4M)]\}$$

up to e^2. We can replace $\sqrt{1 - e^2}$ by 1 and the coefficient of $\partial R/\partial e$ in the Lagrange equation by $1/na^2 e$. Multiplication leaves only terms in e^{-1} and constants. We have

$$\frac{\sqrt{1 - e^2}}{na^2 e}\frac{\partial R}{\partial e} = \frac{\mu J_2}{na^5}\left\{(-\tfrac{1}{4} + \tfrac{3}{4}\cos^2 i)\left(3 + \frac{3}{e}\cos M + 9\cos 2M\right) + (\tfrac{3}{4} - \tfrac{3}{4}\cos^2 i)\right.$$

$$\times\left[-\frac{1}{2e}\cos(2\omega + M) - 5\cos(2\omega + 2M) + \frac{7}{2e}\cos(2\omega + 3M)\right.$$

$$\left.\left. + 17\cos(2\omega + 4M)\right]\right\}$$

Neglecting e, we similarly obtain

$$\frac{-\cos i}{na^2\sqrt{1-e^2}\sin i}\frac{\partial R}{\partial i} = \frac{\mu J_2}{na^5}[\tfrac{3}{2}\cos^2 i - \tfrac{3}{2}\cos^2 i \cos(2\omega + 2M)]$$

Addition of the two quantities gives

$$\frac{d\omega}{dt} = \frac{\mu J_2}{na^5}\left\{-\tfrac{3}{4} + \tfrac{1.5}{4}\cos^2 i + (-\tfrac{1}{4} + \tfrac{3}{4}\cos^2 i)\right.$$

$$\left(\frac{3}{e}\cos M + 9\cos 2M\right) - \tfrac{3}{2}\cos^2 i \cos(2\omega + 2M) + (\tfrac{3}{4} - \tfrac{3}{4}\cos^2 i)$$

$$\times\left[-\frac{1}{2e}\cos(2\omega + M) - 5\cos(2\omega + 2M)\right.$$

$$\left.\left. + \frac{7}{2e}\cos(2\omega + 3M) + 17\cos(2\omega + 4M)\right]\right\} \tag{30}$$

We substitute

$$a = a_0; \quad e = e_0; \quad i = i_0; \quad \omega = \omega_0$$

and

$$M_0 = n(t - t_0) \quad \text{with} \quad n_0^2 a_0^3 = \mu$$

in the right-hand side of the two-body solution, and then carry out the integration:

$$\omega = \omega_0 + \frac{n_0 J_2}{a_0^2}\left(-\tfrac{3}{4} + \tfrac{1.5}{4}\cos^2 i_0\right) t + \left[\frac{J_2}{a_0^2}\left(-\tfrac{1}{4} + \tfrac{3}{4}\cos^2 i_0\right)\right.$$

$$\left(\frac{3}{e_0}\sin M_0 + \tfrac{9}{2}\sin 2M_0\right)\left(-\tfrac{1.5}{8} + \tfrac{9}{8}\cos^2 i_0\right)\sin(2\omega_0 + 2M_0) + (\tfrac{3}{4} - \tfrac{3}{4}\cos^2 i_0)$$

$$\times\left.\left(-\frac{1}{2e_0}\sin(2\omega_0 + M_0)\right) + \frac{7}{6e_0}\sin(2\omega_0 + 3M_0) + \tfrac{17}{4}\sin(2\omega_0 + 4M_0)\right]$$

We thus find the following expression for the coefficient of time in the secular motion of the perigee:

$$\frac{n_0 J_2}{a_0^2}\left(-\tfrac{3}{4} + \tfrac{1.5}{4}\cos^2 i_0\right) \tag{31}$$

which coincides up to e_0 (which has been neglected here) with the result of the previous section (27). We have thus obtained a number of short-period periodic terms. Some of these include e in the denominator. The consequences of this fact will be discussed later. We shall assume only that e is not cancelled out (it should be borne in mind, however, that we have stopped the expansion at e^4 and then at e merely to simplify the argument, and in actual fact, the results of the first approximation can be extended in e as far as required).

We can similarly express the right-hand sides of other Lagrange equations (cf. section 34), e.g. those giving

$$\frac{da}{dt}, \quad \frac{de}{dt}, \quad \frac{di}{dt} \quad \text{and} \quad \frac{d\Omega}{dt}$$

and can obtain the solution in the first approximation by integration as above.

The equation in dM/dt is to be treated separately, as we have already mentioned in section 51. In fact, we have

$$\frac{dM}{dt} = n - \frac{2}{na} \frac{\partial R}{\partial a} - \frac{1 - e^2}{na^2 e} \frac{\partial R}{\partial e}$$

where $n = \sqrt{\mu a^{-3/2}}$. To obtain all the M terms of the same type as the terms of the other variables (i.e. secular short-period terms of first order in J_2), we must take into account the perturbations of first order in n. We therefore first calculate

$$n = \sqrt{\bar{\mu}}\, \bar{a}^{-3/2}$$

according to the first approximation of a. Neglecting J_2^2, we obtain for a that

$$a = a_0 \left[1 + \frac{J_2}{a_0^2} \sum_{jk} A_{jk} \cos(j\omega + kM) \right]$$

and therefore

$$a^{-3/2} = a_0^{-3/2} \left[1 + \frac{J_2}{a_0^2} \sum_{jk} A_{jk} \cos(j\omega + kM) \right]^{-3/2}$$

$$\sqrt{\bar{\mu}}\, \bar{a}^{3/2} = n_0 - \tfrac{3}{2} n_0 \frac{J_2}{a_0^2} \sum_{jk} A_{jk} \cos(j\omega + kM) + \cdots$$

It follows, therefore, that dM/dt is the sum of the three series:

$$\sqrt{\bar{\mu}}\, \bar{a}^{3/2}; \quad \frac{-2}{na} \frac{\partial R}{\partial a}; \quad \text{and} \quad -\frac{1 - e^2}{na^2 e} \frac{\partial R}{\partial e},$$

all having the same form. After integration, M is thus of the same form as the other five elements.

66. Second Approximation with the Lagrange Equations

To obtain the next approximation, we substitute into the right-hand side of the Lagrange equations the series obtained in the first approximation. We use for this the expanded form, such as that obtained for $d\omega/dt$ (eq. 30).

In these calculations, in which we keep only the terms in J_2 and J_2^2, we obtain the same terms in J_2 as in the first approximation. None of the terms in J_2 contains ω

alone, since the term in cos 2ω was absent in the expansion of the disturbing function (section 59). In the above substitution, however, terms in cos 2ω or sin 2ω can and in fact do appear.

We thus find the following expression for $d\omega/dt$:

$$\frac{d\omega}{dt} = J_2\omega_1' + J_2^2\omega_2' + J_2\sum_{ij} A_{ij}\cos(i\omega + jM) + J_2^2\sum_{ij} B_{ij}\cos(i\omega + jM) \tag{32}$$

where the summation is carried out over all values of j, whilst $i=0$ or 2 in A_{ij} and $i=0, 2$, or 4 in B_{ij}; furthermore $A_{20}=0$. The quantities ω_1', ω_2', A, and B are functions of a_0, e_0, and i_0.

The integration is carried out as in the first approximation, but this time we cannot neglect the secular terms of ω and M that are of the order of J_2, since the result is to be exact to J_2^2. Let $J_2\omega_1'$ and $n+J_2n_1'$ be the coefficients of time in the solution of the first approximation for ω and M.

The integrated form of (32) will then be

$$\omega - \omega_0 = (J_2\omega_1' + J_2^2\omega_2')t + J_2\sum_{ij} A_{ij}\frac{\sin(i\omega + jM)}{iJ_2\omega_1' + j(n + J_2n_1')}$$

$$+ J_2^2\sum_{ij} B_{ij}\frac{\sin(i\omega + jM)}{iJ_2\omega_1' + j(n + J_2n_1')} \tag{33}$$

(a) Since A_{00} and A_{20} do not exist, j is always non-zero in the first denominator, and we can write (to within J_2^2)

$$\frac{1}{iJ_2\omega_1' + jn + jJ_2n_1'} = \frac{1}{jn}\left[1 - \frac{J_2}{n}\left(\frac{i}{j}\omega_1' + n_1'\right)\right]$$

The terms of the first sum are thus transformed into terms of first order in J_2 and other terms in J_2^2 of the same form.

(b) As regards the terms of the second form, except those in $B_{2,0}$ ($B_{0,0}$ and $B_{4,0}$ are zero), we can neglect J_2 in the denominator, and integrate as if ω were constant and

$$M = n(t - t_0)$$

In the term in $B_{2,0}$ on the other hand, the divisor becomes $2J_2\omega_1'$ and after integration, the term becomes

$$\frac{J_2^2 B_{20}}{2J_2\omega_1'}\cos 2\omega = \frac{J_2 B_{20}}{2\omega_1'}\cos 2\omega$$

which is a first-order term.

We have thus arrived at an important result which is valid for several problems in celestial mechanics: long-period first-order terms appear only in the second approximation.

It follows, therefore, that the second-order terms calculated in the second approximation are incomplete. They have been calculated with incomplete first-order terms. Consequently, the substitutions whereby we could write the equations up to the second-order form were incomplete. This result is analogous with that formulated in section 52.

Subsequent approximations are carried out similarly: the order of the expansion of the equations and the divisors is increased by one in each step. Thus, at the third approximation, we shall proceed with the expansions up to J_2^3.

The solution will be exact in all the short-period and long-period terms to within J_2^2.

67. Comparison of the Two Methods

Although expressed formally in different forms, the results obtained by the above two methods of solving the same equations should be identical.

The first method, a modification of Delaunay's method, devised by Von Zeipel and Brouwer, gives the results in finite form by defining new quantities r' and v' (cf. section 61). In the case of more complex problems, this transformation is not always possible, and we would have to carry out the same expansions in powers of eccentricity as in the Lagrange method, in which case, the results would be identical.

Thus, the advantage that the results of the first method are valid irrespective of the eccentricity disappears in more complicated problems. The main difference between the two methods is as follows: the method of the Lagrange equations is essentially one of successive approximations, in which we repeat at each step the same calculations with longer and longer expansions. On the other hand, we can proceed with the Von Zeipel method up to any required accuracy, the only condition being that the identities at higher orders of J_2 are satisfied.

The Lagrange method offers a means of obtaining with less calculation a first-order theory valid for small eccentricities. On the other hand, the Von Zeipel method is to be favoured when a greater accuracy is required, particularly as regards the long-period terms. The same applies to other problems of celestial mechanics. The Lagrange method gives a quick and rough estimate of the problem and reveals the main features of the motion. For greater accuracy, however, more economical methods must be used. A number of such methods have been developed, each being more or less adapted to a given problem. Some of these methods will be discussed in Chapters VI and VII, but only in principle, since their application is tedious. Comparison between these methods is often difficult, and the reason for the preference for one or another amongst the various schools is not always clear.

68. The Case of Very Small Eccentricity and Inclination

As we have already seen in section 35, the Lagrange equations are not valid when the eccentricity and the inclination in the motion can be zero, since e and $\sin i$ feature in the denominator of the right-hand sides of the Lagrange equations.

This limitation also applies to the Von Zeipel method. Thus, in the definition of the variable g' by (17) with the aid of S_1 in (21) [section 61]:

$$S_1 = \frac{\mu^2 J_2}{G'^3}\left[\left(-\frac{1}{4}+\frac{3}{4}\frac{H'^2}{G'^2}\right)(v'-l+e'\sin v') + \left(\frac{3}{4}-\frac{3}{4}\frac{H'^2}{G'^2}\right)\frac{1}{2}\sin(2g+2v')\right.$$
$$\left. +\frac{e'}{2}\sin(2g+v') + \frac{e'}{6}\sin(2g+3v')\right]$$

(34)

we have that

$$g' = g + \frac{\partial S_1}{\partial G'} + \frac{\partial S_1}{\partial e'}\frac{\partial e'}{\partial G'} + \frac{\partial S_1}{\partial v'}\frac{\partial v'}{\partial e'}\frac{\partial e'}{\partial G'}$$

since S_1 depends on G' also through

$$e' = \sqrt{1-\frac{G'^2}{L'^2}}$$

and on v' connected with l by

$$E' - e'\sin E' = l; \quad \tan\frac{v'}{2} = \sqrt{\frac{1+e'}{1-e'}}\tan\frac{E'}{2}$$

We have that $\partial e'/\partial G' = -G'/L'^2 e'$; this introduces e' in the denominator, which is not eliminated during the calculations.

It follows therefore that the method described here is not valid when the eccentricity is very small.

The situation is the same as regards zero inclination, since $\tan i$ appears in the denominator of the long-period inclination terms. In these cases, therefore, we must use the variables proposed in section 33.

However, since the perturbations of the inclination are small, a good solution can be obtained by treating the case as planar, in which case, the variables i and Ω are ignored. As regards the eccentricity, however, we shall have to use the variables η, θ, and λ (section 33), unless we employ a completely different method specially adapted to the case of zero mean eccentricity (e.g. Hill's method, see Chapter VI).

69. Critical Inclination

We have seen in section 66 that, in the second approximation by the Lagrange method, we may find long-period terms whose denominator contains $J_2\omega'_1$, i.e. the mean motion of the perigee.

According to (31), we have that

$$J_2\omega'_1 = \frac{n_0 J_2}{a_0^2}\left(-\tfrac{3}{4}+\tfrac{15}{4}\cos^2 i_0\right)$$

Thus, the divisor of the long-period terms is proportional to $1-5\cos^2 i_0$ and tends to zero as $\cos^2 i_0$ tends to $\frac{1}{5}$ (this critical inclination i_0 is about $63°26'$). Thus, the amplitude of these terms increases indefinitely, as does the period. The methods described above, in which J_2 is considered as the only small quantity, lose their validity when $(1 - 5 \cos^2 i)$ is comparable with J_2. We must therefore use a different technique. We shall show how the Von Zeipel method can be adapted to this case. We shall take up the problem after the elimination of l (sections 61 and 62). The task is to solve the canonical set of equations (14) whose Hamiltonian is $\Phi' = \Phi'_0 + \Phi'_1 + \Phi'_{2S} + \Phi'_{2p}$, as given by (18), (19), and (22).

We include terms in J_4, and obtain, in addition to Φ'_{2S}:

$$\Phi^*_{2S} = \Phi'_{2S} + \frac{\mu^6 J_4}{L'^{10}} \left(\frac{15}{16} \frac{L'^7}{G'^7} - \frac{9}{16} \frac{L'^5}{G'^5} \right) \left(\frac{3}{8} - \frac{15}{4} \frac{H'^2}{G'^2} + \frac{35}{8} \frac{H'^4}{G'^4} \right)$$

$$\Phi^*_{2p} = \Phi'_{2p} + \frac{\mu^6 J_4}{L'^{10}} \left(\frac{L'^5}{G'^5} - \frac{L'^7}{G'^7} \right) \left(\frac{15}{64} - \frac{15}{8} \frac{H'^2}{G'^2} + \frac{105}{64} \frac{H'^4}{G'^4} \right) \cos 2g'$$

We shall now attempt to calculate a new determining function:

$$S' = L''l' + G''g' + H''h' + \sum_n S'_n(L'', G'', H'', g') \tag{35}$$

where L'', G'', H'', l'', g'', and h'' are the new variables. The index n, to be defined later, characterizes the terms of different orders in S'.

In the general calculation in section 63, the first S'_n term (S'_1) is given by (25); we note that the numerator includes J_2 and the denominator $1 - 5(H''^2/G''^2) = 1 - 5 \cos^2 i_0$.

If, therefore, $1 - 5(H''^2/G''^2)$ is infinitely small of order λ in J_2, we can no longer say that S'_1 is of the first order. We shall denote its order by v ($v < 1$ and $\lambda < 1$).

We now apply the method in section 63, take into account (35), and keep the first term S'_v of this expansion.

We denote by $\Phi'' = \Phi''_0 + \Phi''_1 + \Phi''_2 + \dots$ the new Hamiltonian, which is independent of l'', g'', and h'', and equate it with Φ'; this procedure gives (up to the second order):

$$\Phi''_0 + \Phi'' + \Phi''_2 = \Phi'_0 + \Phi'_1 \left(L'', G'' + \frac{\partial S'}{\partial g'}, H'' \right) + \Phi^*_{2S} + \Phi^*_{2p} \tag{36}$$

where

$$\Phi'_1 \left(L'', G'' + \frac{\partial S'}{\partial g'}, H'' \right) = \Phi'_1 (L'', G'', H'') + \frac{\partial \Phi'_1}{\partial G''} \frac{\partial S'_v}{\partial g'} + \frac{1}{2} \frac{\partial^2 \Phi'_1}{\partial G''^2} \left(\frac{\partial S'_v}{\partial g'} \right)^2 + \dots$$

and

$$\frac{\partial \Phi'_1}{\partial G''} = \frac{\mu^4 J_2}{L''^3 G''^4} \left(\frac{3}{4} - \frac{15}{4} \frac{H''^2}{G''^2} \right)$$

The zero-order terms are always Φ''_0 and Φ'_0 those of the first order being Φ''_1 and Φ'_1. We thus have that

$$\Phi''_0 = \Phi'_0; \qquad \Phi''_1 = \Phi'_1$$

Terms of a higher order must be second order-terms, i.e. Φ''_2, Φ'_{2s}, and Φ'_{2p}, but we shall also have

$$\frac{\partial \Phi'_1}{\partial G''} \frac{\partial S'_v}{\partial g'} \quad \text{of the order of} \quad 1 + \lambda + v$$

and

$$\frac{\partial^2 \Phi'_1}{\partial G''^2}\left(\frac{\partial^2 S'_v}{\partial g'^2}\right)^2 \quad \text{of the order of} \quad 1 + 2v$$

since $\partial^2 \Phi'/\partial G''^2$ does not contain $(1 - 5[H''^2/G''^2])$ as a factor.

If $(1 + \lambda + v) \neq 2$, we must equate the first terms independently of the second-order terms, and then $\partial S'_v/\partial g' = 0$, S' is independent of g', and $G' = G''$, i.e. the change of variables retains the identity. This, of course, is contrary to our hypotheses. The same applies when $(1 + 2v) \neq 2$, since $\partial S'_v/\partial g'$ is then again zero. We therefore take simultaneously

$$1 + \lambda + v = 2$$

$$1 + 2v = 2$$

and hence $v = \lambda = \frac{1}{2}$.

We can therefore carry out the calculation with a nearly critical inclination, by expanding S' in such a way that the first term is of order $\frac{1}{2}$ in J_2. In fact, the following calculation remains valid even when $(1 - 5[H''^2/G''^2])$ is smaller.

Thus, when we put

$$\Phi^*_{2p} = Q \cos 2g' = 2Q \cos^2 g' - Q$$

the equating of the second-order terms in (36) gives

$$\Phi''_2 = \Phi^*_{2s} + 2Q \cos^2 g' - Q +$$
$$\frac{\partial}{\partial G''}(\Phi'_1 + \Phi^*_{2s} + 2Q \cos^2 g' - Q)\frac{\partial S'_{1/2}}{\partial g'} + \frac{1}{2}\frac{\partial^2 \Phi'_1}{\partial G''^2}\left(\frac{\partial S'_{1/2}}{\partial g'}\right)^2$$

where $S'_{\frac{1}{4}}$ depends on g'. Being of the second-order, the terms $\partial \Phi^*_{2s}/\partial G''$ and $\partial Q/\partial G''$ are smaller and should be neglected in the next approximation. However, to retain the secular terms which do not tend to zero when $i \to i_0$, we shall neglect only $\partial(2Q \cos^2 g)/\partial G''$ and retain the identity

$$\Phi''_2 = \Phi^*_{2s} - Q + 2Q \cos^2 g' + \frac{\partial}{\partial G''}(\Phi'_1 + \Phi^*_{2s} - Q)\frac{\partial S'_{1/2}}{\partial g'} + \frac{1}{2}\frac{\partial^2 \Phi'_1}{\partial G''^2}\left(\frac{\partial S'_{1/2}}{\partial g'}\right)^2 \qquad (37)$$

Separation of the terms independent of g' leads to the two equalities:

$$\left\{ \begin{array}{l} \Phi''_2 = \Phi^*_{2s} - Q \\ \frac{1}{2}\frac{\partial^2 \Phi'_1}{\partial G''^2}\left(\frac{\partial S'_{1/2}}{\partial g'}\right)^2 + \frac{\partial}{\partial G''}(\Phi'_1 + \Phi^*_{2s} - Q)\frac{\partial S'_{1/2}}{\partial g'} + 2Q \cos^2 g' = 0 \end{array} \right\} \qquad (38)$$

The first of these gives Φ_2'', whilst the second defines $\partial S_4'/\partial g'$ and thus S_4'. Putting

$$A = \frac{-\partial\,(\Phi_1' + \Phi_{23}^* - Q)/\partial G''}{\partial^2\Phi_1'/\partial G''^2}$$

$$B = \frac{4Q}{\partial^2\Phi_1'/\partial G''^2} \quad \text{and} \quad g^* = g' - \frac{\pi}{2}$$

we obtain

$$\frac{\partial S_{1/2}'}{\partial g^*} = A \pm \sqrt{A^2 - B \sin^2 g^*} \tag{39}$$

70. Libration of the Perigee near the Critical Inclination

The integral of (39) cannot be expressed with the aid of elementary functions, and we must introduce elliptical functions.

If we replace H'^2/G''^2 in B by $\frac{1}{3}$, we find that, in the vicinity of critical inclination, we have

$$B = \frac{J_2\mu^2}{5L'^2}\left(1 - \frac{L'^2}{G''^2}\right)\left(1 - \frac{J_4}{J_2^2}\right)$$

The sign of B thus depends on the sign of $(J_4 - J_2^2)$. Since $G''^2 = L'^2(1 - e'^2) < L'^2$, we have that

$$B < 0 \quad \text{if} \quad J_4 < J_2^2$$
$$B = 0 \quad \text{if} \quad J_4 = J_2^2$$
$$B > 0 \quad \text{if} \quad J_4 > J_2^2$$

In the case of the Earth, we have the approximate values:

$$J_2 = 0{,}00108; \quad J_4 = 0{,}000002$$

$$\frac{J_4}{J_2^2} \neq 1{,}7 \quad \text{thus} \quad B > 0$$

We shall now examine this case and put $k^2 = B/A^2$. Integration of (39) gives

$$S_{1/2}' = Ag^* - A\int_0^{g^*}\sqrt{1 - k^2 \sin^2 g^*}\,dg^*$$

Let us examine more closely the motion of the perigee. The equations of the form (23), which give the change of variables with the aid of the determining function (35), enable us to write

$$G'' = G' - \frac{\partial S_{1/2}'}{\partial g^*}; \quad g_1'' = g^* + \frac{\partial S_{1/2}'}{\partial G''} \tag{40}$$

with $g''_1 = g'' - \pi/2$, so as to take into account the change of variable $g^* = g' - \pi/2$ made above.

The solution being of the form of

$$G'' = \text{constant} \quad \text{and} \quad g''_1 = \left(-\frac{\partial \Phi}{\partial G''} \right)(t - t_0) \tag{41}$$

equation (40) in g''_1 is written as

$$g''_1 = \left(1 + \frac{\partial A}{\partial G''} \right) g^* - \left(\frac{\partial A}{\partial G''} - \frac{A}{2B} \frac{\partial B}{\partial G''} \right) \int_0^{g^*} \frac{d\varphi}{\sqrt{1 - k^2 \sin^2 \varphi}}$$

$$- \frac{A}{2B} \frac{\partial B}{\partial G''} \int_0^{g^*} \sqrt{1 - k^2 \sin^2 \varphi} \, d\varphi \tag{42}$$

and is an implicit function of g^* of the argument g''_1, which is itself a linear function of time.

By the definition of the critical inclination i_0, the quantity $k = \sqrt{B/A}$ tends to infinity as $i \to i_0 (A \to 0)$. In the vicinity of $i = i_0$ we see that $k > 1$ and we must change the variables defined by $\tan \varphi = \cos \theta / \sqrt{k^2 - 1}$ in (42) to find the classical elliptical functions.

This calculation requires manipulating with elliptical functions and, instead of reproducing it here, we shall merely state that, as long as A is small $(k^2 > 1)$ the inverse of the function defined by (42) is periodic with a period of

$$P = 4 \int_0^{\text{arc sin} \frac{1}{k}} \frac{d\theta}{\sqrt{1 - k^2 \sin^2 \theta}}$$

and with an amplitude arc sin $1/k$.

When $1/k = 0$, $A = 0$, the solution reduces to $g^* = 0$, $g'' = \pi/2$, and the perigee is fixed at a point whose latitude is the highest.

When $1/k \neq 0$, but $k > 1$, the above considerations show that the perigee performs a periodic motion about the position $g^* = 0$, with an amplitude $1/k$ and a period P. We note that this period depends essentially on the initial conditions, and is in fact particularly sensitive to these. In actual fact, $1/k$ varies from 0 to 1 when the inclination of the orbit of a satellite fairly close to the Earth varies only from i_0 to $i_0 \pm 8'$.

This period lacks the stability of the periods discussed so far (periods of rotation), which vary slowly and continuously with the initial conditions. The new type of period is called the period of *libration*, and will be described in section 71.

As k tends to 1, the period increases indefinitely with an amplitude tending to 90°. At the limit, we have an asymptotic motion of the perigee towards one of the nodes, this motion having an infinite duration.

When $k < 1$, the perigee revolves around the Earth. The period introduced is the period of this revolution. This is the general case treated in the present Chapter. Although (42) remains valid and it describes the motion, the solutions of sections 64 and 65 represent its equivalents and are also valid. However, identification of these two expressions is difficult, and convergence is better in one or the other depending on whether the motion is asymptotic ($k \simeq 1$) or not (k small).

71. The Phenomenon of Libration

Libration is a common phenomenon in celestial mechanics and can be observed in the motion of a number of bodies in the solar system. A well-known example is represented by the Trojan asteroids which are 60° ahead and 60° behind Jupiter in its orbit. Certain satellites, such as Hyperion, also exhibit this type of motion. Finally, the rotation of the Moon about its own axis is of this type (lunar libration), although it is not associated with orbital motion.

Libration appears every time the solution contains long-period terms whose period tends to infinity under certain initial conditions. This period may be that of the argument of the perigee, as in the example above. However, it is more often the period of resonance, derived from the fact that two of the periods in the arguments of the perturbation function are commensurate. Any long-period term can give rise to libration in an adjacent motion, provided that the equilibrium position corresponding to perfect resonance is stable. The essential property of libration is that its period, which replaces the period of revolution in the libration domain, is very sensitive to small variations in the initial conditions. Moreover, the initial conditions giving rise to libration are generally comprised within narrow limits, the libration domain bordering on that of asymptotic motion.

The types of motion around a libration domain differ greatly, and certain properties of neighbouring trajectories may be discontinuous (e.g. those of the perhelions). However, these motions can be likened to that of a simple pendulum. This similarity is also exhibited in the equations. According to the initial conditions (impulses), the motion of a simple pendulum may be pendular (periodic, to and fro), asymptotic (tending towards the unstable higher equilibrium position), or of the type of revolution (pendulum turning around its axis).

A very small variation in the initial conditions may completely change the conditions of the motion. The same applies in celestial mechanics, and study of these features of motion is one of the most difficult aspects of this branch of science.

CHAPTER VI

LUNAR THEORY AND THE MOTION OF THE SATELLITES

Since the Moon is the nearest celestial body, its position has been observed with the greatest accuracy. It is therefore natural that the problem of its motion should be solved with a refinement that cannot be expected in the case of other celestial bodies. This has been attempted in various ways in the last two centuries by a great number of mathematicians whose work has made a major contribution to celestial mechanics and made the problem of the motion of the Moon (lunar theory) one of the key problems of this science. The outstanding names in this connection are those of Laplace, Ponté-coulant, Hansen, Delaunay, Hill, and Brown.

72. The Principal Problems of the Lunar Theory

The revolution of the Moon around the Earth is perturbed mainly by the Sun. Other celestial bodies also cause some perturbation, but, as we have seen in section 8, their action is much weaker. The oblate shape of the Earth (see section 55) also has an effect, but, on account of the distance between the Earth and the Moon, this effect is very small (section 56). In fact, the operative perturbations are well approximated by assuming that the Sun is the only perturbing body and that the Earth revolves around it in a fixed Keplerian ellipse. Study of the motion under these simplified conditions is known as *the main problem of the lunar theory*.

Perturbations due to the planets are called *direct planatery perturbations*, whilst the *indirect planetary perturbations* are those differences in the motion of the Moon which are caused by the fact that, owing to the planetary perturbations suffered by the Earth, the latter's orbit is not a perfect ellipse. The indirect planetary perturbations are stronger than the direct ones, but much weaker than the solar perturbations.

To formulate the motion of the Moon we calculate either the variations of the osculating elements or the variations of the spherical coordinates. These variations are expressed as the sum of periodic terms called *inequalities*. Some of these have been known for a long time: evection was already known to Hipparchus and variation to Kepler, not to mention the retrograde motion of the nodes and the advance of the perigee, whose periods together with the lunar month determine the recurrence of eclipses. We shall first discuss briefly the principal inequalities, and then describe the foundations underlying the best-known theories of the Moon.

73. Approximate Solution of the Main Problem

The disturbing function of this problem was constructed in sections 45 and 46, and expressed (eq. 41, section 46) as a function of the osculating variables a, i, Ω, ω, quantities r and v (radius vector and the true anomaly), and the analogous quantities r', and v' pertaining to the Sun relative to the Earth.

The following simplifications are introduced in the approximate study of the principal inequalities:

(a) Since the inclination i of the Moon is small ($5°8'$), we can neglect i^4 in the disturbing function.

(b) We can neglect the eccentricity e' of the Earth's orbit ($e'=0.016$).

(c) We can neglect the terms higher than e^2, where e is the eccentricity of the Moon's orbit ($e=0.054$).

It should be added, however, that these simplifications are not used in the exact study of the motion of the Moon.

Using point (a) above, we can replace $\sin i$ by i and $\cos^2 i/2$ by $(1-i^2)$ in the perturbation function (eq. 41 in Chapter IV). Furthermore, point (b) enables us to replace a'/r' by 1, and v' by the mean anomaly M' of the Sun. We thus have that

$$R_1 = n'^2 a^2 \left(\frac{r}{a}\right)^2 \left[\tfrac{1}{4} - \tfrac{3}{8}i^2 + \tfrac{3}{4}\left(1 - \frac{i^2}{2}\right) \cos 2(\omega - \omega' + v - M' + \Omega) \right.$$
$$\left. + \frac{3i^2}{8}\left(\cos(2\omega + 2v) + \cos(2\omega' + 2M' - 2\Omega)\right) \right] \tag{1}$$

The use of point (c) and a calculation similar to that of section 40 on the expansion of the two-body functions leads to

$$\left(\frac{r}{a}\right)^2 \cos 2v = \tfrac{5}{2}e^2 - 3e \cos M + (1 - \tfrac{5}{2}e^2) \cos 2M + e \cos 3M + e^2 \cos 4M \tag{2}$$

$$\left(\frac{r}{a}\right)^2 \sin 2v = - 3e \sin M + (1 - \tfrac{5}{2}e^2) \sin 2M + e \sin 3M + e^2 \sin 4M \tag{3}$$

$$\left(\frac{r}{a}\right)^2 = 1 + \frac{3e^2}{2} - 2e \cos M - \frac{e^2}{2} \cos 2M \tag{4}$$

The identity of the coefficients of the first two expansions is no longer preserved when the expansions in e are carried further. The substitution is carried out with terms which do not contain i^2 as a factor. Owing to the smallness of i^2, we neglect the terms in $i^2 e$ and, a fortiori, those in $i^2 e^2$.

We can thus replace $(r/a)^2$ by 1, and v by M whenever i^2 is a factor. We replace

$$\left(\frac{r}{a}\right)^2 \cos 2(\omega - \omega' + v - M' + \Omega)$$

by

$$\left(\frac{r}{a}\right)^2 \cos 2v \cos 2(\omega - \omega' - M' + \Omega) - \left(\frac{r}{a}\right)^2 \sin 2v \sin 2(\omega - \omega' - M' + \Omega)$$

Combining this with (2) and (3), we obtain

$$\tfrac{5}{2} e^2 \cos 2(\omega - \omega' - M' + \Omega) - 3e \cos(2\omega - 2\omega' - 2M' + 2\Omega + M)$$
$$+ (1 - \tfrac{5}{2} e^2) \cos 2(\omega - \omega' - M' + \Omega + M)$$
$$+ e \cos(2\omega - 2\omega' - 2M' + 2\Omega + 3M)$$
$$+ e^2 \cos(2\omega - 2\omega' - 2M' + 2\Omega + 4M)$$

We thus have the following expansion of R_1:

$$
\begin{aligned}
R_1 = n'^2 a^2 \Big[& \frac{1}{4} + \frac{3e^2}{8} - \frac{e}{2} \cos M - \frac{e^2}{8} \cos 2M \\
& + \tfrac{15}{8} e^2 \cos(2\omega - 2\omega' - 2M' + 2\Omega) \\
& - \frac{9}{4} e \cos(2\omega - 2\omega' - 2M' + 2\Omega + M) \\
& + (\tfrac{3}{4} - \tfrac{15}{8} e^2) \cos(2\omega - 2\omega' - 2M' + 2\Omega + 2M) \\
& + \tfrac{3}{4} e \cos(2\omega - 2\omega' - 2M' + 2\Omega + 3M) \\
& + \frac{3e^2}{4} \cos(2\omega - 2\omega' - 2M' + 2\Omega + 4M) \Big] \\
& + n'^2 a^2 i^2 [-\tfrac{3}{8} - \tfrac{3}{16} \cos(2\omega - 2\omega' - 2M' + 2\Omega + 2M) \\
& + \tfrac{3}{8} \cos(2\omega + 2M) + \tfrac{3}{8} \cos(2\omega' + 2M' - 2\Omega)]
\end{aligned}
\tag{5}
$$

We can apply the Lagrange equations to this disturbing function, and thus obtain the variations of the elements.

Thus, the non-periodic part of R_1

$$R_s = n'^2 a^2 \left(\tfrac{1}{4} + \frac{3}{8} e^2 - \tfrac{3}{8} i^2 \right)$$

enables us to calculate in the first approximation (cf. section 51, a) the secular motion of the perigee and the nodes by the use of the Lagrange formulae [eq. (45), Chapter III]:

$$\frac{d\Omega}{dt} = \frac{1}{na^2 \sqrt{1 - e^2} \sin i} \frac{\partial R_s}{\partial i} = \frac{n'^2 a^2}{na^2 i} \left(1 + \frac{e^2}{2} \right) \left(-\frac{6}{8} i \right) = -\frac{3}{4} \frac{n'^2}{n} \left(1 + \frac{e^2}{2} \right)$$

$$\frac{d\bar{\omega}}{dt} = \frac{\sqrt{1 - e^2}}{na^2 e} \frac{\partial R_s}{\partial e} - \frac{\cos i}{na^2 \sqrt{1 - e^2} \sin i} \frac{\partial R_s}{\partial i}$$

$$= \frac{n'^2 a^2}{na^2 e} \left(1 - \frac{e^2}{2} \right) \frac{6e}{8} + \frac{n'^2 a^2}{na^2 i} \left(1 + \frac{e^2}{2} \right) \frac{6i}{8} = \frac{3}{2} \frac{n'^2}{n}$$

We obtain from this that

$$\frac{d\bar{\omega}}{dt} = \frac{d\Omega}{dt} + \frac{d\bar{\omega}}{dt} = \frac{3}{4} \frac{n'^2}{n} \left(1 - \frac{e^2}{2} \right)$$

The following is a more exact expression (e is neglected):

$$\frac{d\varpi}{dt} = n\left[\frac{3}{4}\left(\frac{n'}{n}\right)^2 + \frac{225}{32}\left(\frac{n'}{n}\right)^3 + \frac{4071}{128}\left(\frac{n'}{n}\right)^4 + \cdots\right]$$

NUMERICAL APPLICATION

The mean motion of the Sun is $n' = 360°$ in a year of 365.25 days, whilst that of the Moon for the same period is $n = 4812.7°$.

We thus find that $d\bar{\Omega}/dt = -20.16°$ per annum and $d\bar{\omega}/dt = 39.09°$, which give periods of 17.86 and 9.21 years respectively. The corresponding values in the complete theory of the real motion are 18.60 and 8.85 years.

74. The Principal Inequalities of the Motion of the Moon

The effects of the other terms of the disturbing function can be obtained in an analogous manner. However, instead of the osculating elements, we examine the variations in the longitude and the latitude of the Moon. For example, the true longitude is

$$\psi = v + \omega + \Omega = \omega + \Omega + M + \left(2e - \frac{e^3}{4} + \cdots\right)\sin M + (\tfrac{5}{4}e^2 + \cdots)\sin 2M + \cdots \quad (6)$$

This is found from (35) (see section 43). Similarly the latitude φ is given by

$$\sin \varphi = \sin i \sin(\omega + v) \quad (7)$$

where v will also be expressed as a function of e and M.

We substitute the disturbing function R (eq. 5) into the Lagrange equations, carry out an integration, and substitute the results into (6) or (7). It should be noted, however, that, as in section 51, the calculation of M necessitates preliminary integration of the equation in a. The results are trigonometric expressions whose resultant gives the true trajectory of the Moon.

The equation of the centre is essentially the expansion of $v - M$ in lontigude, of $r - a$ for the radius vector, and of the corresponding terms in latitude. These expansions are analogous to those in Chapter IV, to which we should have added the perturbations of the same argument. The equation of the centre describes the basic orbit, which is a slightly deformed ellipse with an eccentricity of 0.0549. This basic orbit performs a double rotation about its focus in its plane (motion of the perigee) and about the axis of the ecliptic plane (motion of the nodes).

The inequalities are mobile deformations of the basic orbit, revolving with variable periods. They are due to various terms in the disturbing function. The principal inequalities are as follows:

1. *The principal perturbation in the latitude*, originating from the R_1 term

$$\tfrac{3}{8}n'^2a^2i^2 \cos(2\omega' + 2M' - 2\Omega) \quad (8)$$

Owing to absence of M in its argument, this gives a divisor of $2n'$, as a result of which the corresponding contribution is increased by a factor of 12. In latitude, this gives a variation proportional to $\sin(\omega + M + 2\Omega - 2\omega' - 2M')$ with a period of 32.28 days and an amplitude of $10'24''$ (i.e., about 550 km).

2. *The evection*, originating mainly from the following R terms:

$$\tfrac{15}{8} n'^2 a^2 e^2 \cos[2(\omega + \Omega) - 2(M' + \omega')]$$
$$- \tfrac{9}{4} n'^2 a^2 e \cos[M + 2(\omega + \Omega) - 2(M' + \omega')] \qquad (9)$$

The calculations give the following result for the evection in longitude

$$1°,274 \sin[M + 2(\omega + \Omega) - 2(M' + \omega')]$$

the period being 31.812 days. This is the most important inequality in the lunar theory. The evection was already known to Hipparchus. It causes a displacement of the Moon by a distance which is nearly 2.5 times its diameter.

3. *The variation*, with an argument of

$$2(M + \omega + \Omega - M' - \omega'),$$

and a period of about 14.78 days (exactly half a lunar month). The corresponding coefficient for the longitude contains

$$\frac{75}{16} e^2 \frac{n'}{n} + \left(\frac{11}{8} + \frac{1101}{64} e^2\right)\left(\frac{n'}{n}\right)^2 + \frac{59}{12}\left(\frac{n'}{n}\right)^3 + \frac{893}{72}\left(\frac{n'}{n}\right)^4 + \cdots \qquad (10)$$

which gives an amplitude of $39'30''$. This inequality was discovered by Tycho Brahé, and it is second only to evection in importance.

4. *The parallactic inequality*, whose period is twice that of the variation, and the coefficient in longitude of its argument $(M + \omega + \Omega - M' - \omega')$ is given by

$$\frac{a}{a'}\left(-\frac{15}{8}\frac{n'}{n} - \frac{93}{8}\left(\frac{n'}{n}\right)^2 + \cdots\right)$$

the amplitude being about $125''$.

It is the greatest inequality having a/a' as a factor, and a comparison of this amplitude with observations that enables one to measure the distance (or the parallax) between the Earth and the Sun as a function of that of the Moon (hence the name "parallactic inequality").

5. *The annual equation*, having a period of one year (argument M') and an amplitude of $668''$. The coefficient of the term in longitude is

$$- 3e' \frac{n'}{n} + \frac{735}{16} e'\left(\frac{n'}{n}\right)^3 + \frac{1261}{4} e'\left(\frac{n'}{n}\right)^4 + \cdots \qquad (11)$$

This depends of course on the eccentricity e' of the Earth's orbit.

There are numerous other inequalities which have not been named. In fact, we find 13 inequalities of longitude, with an amplitude exceeding 100″, and 46 others whose amplitude is between 1 and 100″.

In Brown's theory, which is currently used in the ephemerides, we find 310 inequalities with different periods.

75. Various Lunar Theories

The calculation of all the inequalities with amplitudes within certain limits, carried out in such a manner that they give the observed trajectory, is called *a lunar theory*, and a great number of methods have been used for this calculation. The assumptions made in this theory, which have already been mentioned, refer to the solution of the principal problem. The result is a numerical expression as a function·of time in the geocentric coordinates of the Moon: longitude, latitude, and the radius vector.

The various methods differ mainly in the order of treating the difficulties, in the system of coordinates used, and in the selected orbit of comparison (called an intermediate orbit).

We shall now briefly describe three different theories, put forward by Delaunay, Brown, and Hansen, which have served as the basis of calculations for the, past hundred years. Each of these theories is the outcome of long years of work; we cannot, of course, follow here the thread of the argument, and shall give only a brief description of the characteristic features in each case.

76. Delaunay's Theory

Ch. Delaunay published in 1860 and 1867 his theory representing the most extensive analytical investigation of the problem. His aim was to express the coordinates of the Moon in the form of a Fourier series whose coefficients are finite expansions in the following quantities, which are assumed to be small:

e and e'	eccentricities of the orbits of the Moon and the Sun
$\gamma = \sin i/2$	where i is the inclination
n'/n	ratio of the mean motions of the Sun and the Moon
a/a'	ratio of the semi-major axes.

As in all the problems encountered so far, we start with the orbit of the two-body problem. We expand the perturbation function, and write the equations as a function of the Delaunay variables L, G, H, l, g, and h (cf. section 32), and as a function of the mean anomaly of the Sun $l' = n'(t-t_0)$. The equations are canonical, but the Hamiltonian depends on time. We can make the Hamiltonian independent of time by introducing two new variables: $k = l'$ and a conjugated variable K (cf. section 20). We thus have a canonical system with four pairs of conjugated variables, the new Hamiltonian being obtained from the previous one by the addition of K.

Delaunay separated from the Hamiltonian one important periodic term

$$\Phi = A + B \cos \theta + \Phi_1$$

where A is a term independent of the variables g, h, k, l, and where θ is a linear function with integer values of these variables as its coefficients. $\theta = a_1 g + a_2 h + a_3 k + a_4 l$; Φ_1 contains the rest of the disturbing function.

By changing the canonical variables, we can make θ into one of the canonical variables that we shall call l and obtain the following expression:

$$\Phi = A(L, G, H, K) + B(L, G, H, K) \cos l + \Phi_1$$

The following set of canonical equations with the Hamiltonian $\Phi^* = A + B \cos l$ is solved:

$$\left. \begin{array}{llll}
\dfrac{dL}{dt} = \dfrac{\partial \Phi^*}{\partial l}; & \dfrac{dG}{dt} = \dfrac{\partial \Phi^*}{\partial g}; & \dfrac{dH}{dt} = \dfrac{\partial \Phi^*}{\partial h}; & \dfrac{dK}{dt} = \dfrac{\partial \Phi^*}{\partial k} \\[3mm]
\dfrac{dl}{dt} = -\dfrac{\partial \Phi^*}{\partial L}; & \dfrac{dg}{dt} = -\dfrac{\partial \Phi^*}{\partial G}; & \dfrac{dh}{dt} = -\dfrac{\partial \Phi^*}{\partial H}; & \dfrac{dk}{dt} = -\dfrac{\partial \Phi^*}{\partial K}
\end{array} \right\} \quad (12)$$

Since Φ^* does not depend on g, h, and k, the right-hand side of the last three equations in the first row is zero. Furthermore, G, H, and K are constants in the solution, and we thus need to solve only

$$\dfrac{dL}{dt} = \dfrac{\partial \Phi^*}{\partial l}; \quad \dfrac{dl}{dt} = -\dfrac{\partial \Phi^*}{\partial L}$$

This set can be solved in various ways. The method used by Delaunay is too long to be reproduced here. It amounts to another change of the canonical variables, starting from a determining function:

$$S = L'l + G'g + H'h + K'k + S_1 + S_2 + \cdots$$

exactly analogous with that used in section 63, the problem being in fact the same. We consider only the first-order solution (i.e. only S_1). The new variables $L', G,' H', K', l', g', h'$, and k' are so defined that the new Hamiltonian Φ^* no longer involves l'.

The change of variables thus defined is then applied to the initial set with the Hamiltonian

$$\Phi = \Phi^* + \Phi_1$$

The change of variables makes the term $B \cos l$ disappear from Φ, or at least it reduces by one unit its order in the small quantity characteristic of the perturbing force.

We now carry out the same calculation for a new term of Φ_1, eliminate it in turn, and continue this process until we have eliminated all the terms that might noticeably affect the solution.

When this is done, after N such transformations, we obtain an N-th set of unknowns:

$$L_N, G_N, H_N, K_N, l_N, g_N, h_N \quad \text{and} \quad k_N$$

and a Hamiltonian

$$\Phi_N = A_N(L_N, G_N, H_N, K_N)$$

The equations are now in the following form:

$$\frac{dL_N}{dt} = \frac{\partial \Phi_N}{\partial l_N} = 0; \quad \frac{dG_N}{dt} = \frac{\partial \Phi_N}{\partial g_N} = 0; \quad \frac{dH_N}{dt} = \frac{\partial \Phi_N}{\partial h_N} = 0; \quad \frac{dK_N}{dt} = \frac{\partial \Phi_N}{\partial k_N} = 0$$

$$\frac{dl_N}{dt} = -\frac{\partial \Phi_N}{\partial L_N}; \quad \frac{dg_N}{dt} = -\frac{\partial \Phi_N}{\partial G_N}; \quad \frac{dh_N}{dt} = -\frac{\partial \Phi_N}{\partial H_N}; \quad \frac{dk_N}{dt} = -\frac{\partial \Phi_N}{\partial K_N}$$

We note that L_N, G_N, H_N, and K_N are constants, and l_N, g_N, h_N, and k_N are linear functions of time:

$$l_N = l'_N t - l_0; \quad g_N = g'_N t - g_0; \quad h_N = h'_N t - h_0; \quad k_N = k'_N t - k_0$$

Since we have defined the successive changes of variables, we know the relationships between two successive sets of variables, and can gradually define the relations connecting the initial variables (which are the osculating elements) and the final variables:

$$\left.\begin{aligned} L &= F_L(L_N, G_N, H_N, K_N, l_N, g_N, h_N, k_N) \quad \text{etc.} \\ l &= F_l(L_N, G_N, H_N, K_N, l_N, g_N, h_N, k_N) \quad \text{etc.} \end{aligned}\right\} \tag{13}$$

where L_N, G_N, H_N, and K_N are constants, and the variables l_N, g_N, h_N, and k_N are linear functions of time. We have thus found the solution of the problem. All that remains is to define the arbitrary constants e, γ, a, and n as functions of L_N, G_N, H_N, and K_N in such a way that these constants assume the nature of mean elements or the mean values of the elements for an infinite interval of time. The solution is then in the required form, and enables us to find any function of the osculating elements and particularly the coordinates of celestial bodies.

Thus, in this method, we calculate, by successive elimination of the periodic terms, the perturbations of the elliptical elements equivalent to a system of coordinates. The secular terms appear only at the end of the process.

This method is essentially the same as that of Von Zeipel, described in Chapter V in connection with artificial satellites. In fact, Von Zeipel's method is basically an improved and simplified version of Delaunay's method.

Delaunay treated in this manner more than 230 terms of the disturbing function and obtained the lunar coordinates in nearly 400 terms representing more than 10000 single terms in the limited expansion of these terms as a function of small parameters.

It is now sufficient to replace these parameters by numerical values. The improvement of these numerical values represents merely a substitution in (13). However, the terms omitted by Delaunay are not negligible, and to reach the accuracy of modern observations we would have to carry out the same calculations with at least five times as many terms.

77. The Theory of Hill and Brown

This theory is based on the theoretical work of Hill carried out at the beginning of the present century, and forms the basis of the current ephemerides of the Moon.

One of the main difficulties of Delaunay's theory is the great number of terms to be calculated, owing to the very slow convergence of the series in n'/n. This is the ratio whose numerical value has been determined very accurately from observations. Hill's proposal, further developed and applied by Brown, is that, instead of starting with an elliptical orbit, we should start with an intermediate orbit already containing all the perturbations depending only on n'/n. Another important characteristic is that we use rectangular coordinates instead of the elements, which makes it unnecessary to expand the disturbing function in terms of the elliptical elements.

The intermediate orbit (called variational orbit) is obtained in the following manner: we consider a planar problem. The Moon moves in the plane of the ecliptic, which means that we shall take into account in the solution only the terms that are independent of i. We further suppose that the Earth revolves around the Sun in a circle. This means that we shall calculate in the solution only the terms that are independent of e'. Finally, we shall retain only R_1 from the disturbing function as defined by (39) in Chapter IV:

$$R_1 = \frac{kmr^2}{r'^3}(-\tfrac{1}{2} + \tfrac{3}{2}\cos^2 S); \qquad km = n'^2 a'^3$$

We neglect in the solution a/a', which means that the Sun is at infinity and its attractive force is equal to that exerted at a distance a' (cf. section 73). The equations of motion in planar rectangular coordinates are:

$$\left.\begin{aligned}
\frac{d^2x}{dt^2} &= \frac{-\mu x}{r^3} + \frac{\partial R_1}{\partial x}; \\
\frac{d^2y}{dt^2} &= \frac{-\mu y}{r^3} + \frac{\partial R_1}{\partial y}; \qquad \mu = n^2 a^3
\end{aligned}\right\} \tag{14}$$

Hill chose a set of axes revolving with an angular velocity n'. The axis Tx always passes through the centre of the Earth and is always directed towards the Sun. The circular motion of the latter is exactly the mean motion n' (see Figure 10). We have that

$$X = r \cos S; \quad Y = r \sin S$$
$$X = x \cos n't + y \sin n't$$
$$Y = - x \sin n't + y \cos n't$$

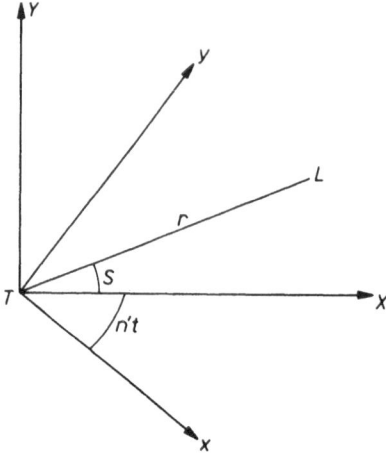

Fig. 10.

This change of variables leads to the following system:

$$\left. \begin{array}{c} \dfrac{d^2X}{dt^2} - 2n'\dfrac{dY}{dt} + \dfrac{\mu X}{r^3} - 3n'^2X = 0 \\[2mm] \dfrac{d^2Y}{dt^2} + 2n'\dfrac{dX}{dt} + \dfrac{\mu Y}{r^3} = 0 \end{array} \right\} \qquad (15)$$

The general solution of the above system gives all the trajectories, solutions of eq. (14), i.e. according to the assumptions made above, the terms in the lunar theory that depend only on n'/n and e.

Using these equations, we attempt to find a solution involving only terms that depend on n'/n alone, i.e. those which correspond to $e=0$. This variational orbit is a periodic solution of the system (15), such that the orbit is symmetrical about the axes OX and OY. The solution is in the following form:

$$\left. \begin{array}{c} X_0 = \displaystyle\sum_{i=0}^{\infty} a_i \cos(2i + 1)(n - n')(t - t_0) \\[3mm] Y_0 = \displaystyle\sum_{i=0}^{\infty} a_i' \sin(2i + 1)(n - n')(t - t_0) \end{array} \right\} \qquad (16)$$

where each of the coefficients a_i and a_i' is a power series in $n'/(n-n')$. We now replace n'/n and consequently the series (16) by numerical values, and thus considerably simplify the calculations. The second step of this method consists in finding a more general solution of (15) so as to obtain the terms depending on e. The required solution is put in the form:

$$X = X_0 + \delta x$$
$$Y = Y_0 + \delta y$$

We substitute in (14), write that X_0 and Y_0 are solutions, put

$$\Omega = \frac{\mu}{r} + \tfrac{3}{2}n_2'X^2$$

and derive the differential equations in δx and δy (their squares are neglected):

$$\frac{d^2\delta x}{dt^2} - 2n'\frac{d\delta y}{dt} = \frac{\partial^2\Omega}{\partial x^2}\delta x + \frac{\partial^2\Omega}{\partial x\partial y}\delta y$$

$$\frac{d^2\delta y}{dt^2} + 2n'\frac{d\delta x}{dt} = \frac{\partial^2\Omega}{\partial x\partial y}\delta x + \frac{\partial^2\Omega}{\partial y^2}\delta y$$

The derivatives of Ω are taken on the variational orbit. The solution of this set involves a second period, which is none other than the period of the perigee.

Using an analogous correction technique, we construct the equation in z (no longer neglecting i), and introduce in the solution a third period: the period of the nodes.

The remainder of the calculation is carried out similarly: we construct the differential equations of the terms to be calculated, noting that the terms calculated previously are solutions of an analogous differential equation.

The various systems obtained are analogous to (15) in form and treatment, other parts of the disturbing function having possibly been added to (15).

Unlike n'/n, the other parameters remain in the algebraic form during the calculation, and are replaced by their numerical values only at the end.

We have seen thus that in this method we first calculate the essential part of the variation, and then the terms depending on e, these terms representing in particular the motion of the perigee and the evection. After this, we calculate the motion of the nodes. The results are then improved in steps and the less important inequalities are included.

Unlike the case in Delaunay's method, the elliptical elements play only a secondary part, since the coordinates are rectangular and the initial intermediate orbit is variational.

78. Hansen's Theory

Published in 1857, this method preceded even that of Delaunay. Hansen gave a solution similar to Delaunay's in precision, but since it was put forward earlier it has been employed more often. Although it was originally formulated for the problem of the Moon, it has been adapted more successfully to problems of the planetary type, and will therefore be discussed briefly in Chapter VII (sections 85–87).

79. Improvement of the Theories

Comparison with more recent observations shows that the above theories, obtained as numerical series representing the coordinates or the osculating elements, need to be improved. The great advantage of a purely algebraic theory, such as that of Delaunay, is that an improvement can be effected simply by the substitution of better numerical values. This cannot be done in the other theories, particularly Brown's, whose improvement seems desirable on the basis of observations made in the past 50 years.

One of the methods used for this purpose is as follows: Consider a set of differential equations in the simplified form

$$\frac{dx_i}{dt} = f_i(x_j) \quad 1 \leqslant i, j \leqslant 6$$

which depends on six unknown x_j.

We assume an approximate solution $\bar{x}_j$ which is to be improved, formed of purely numerical expressions, t being the only litteral quantity. We shall substitute this solution in the right-hand sides of the above equations. Let $f_i(\bar{x}_j)$ be the numerical result of this substitution.

The equations

$$\frac{dx_i^*}{dt} = f_i(\bar{x}_j)$$

are integrated into

$$x_i^* = x_{i0} + \int f_i(\bar{x}_j) \, dt$$

We determine the values of x_{i0}, preferably by comparison with observations, and thus obtain six new x_i^* series. We can repeat this process several times and consider that x_i^* is a new approximate numerical solution to be treated as $\bar{x}_j$ above, and so forth.

Although no general theory of the series of solutions has been developed, it has been shown that this series converges towards the solution even in *difficult* cases, when the parameters n'/n, e, e', and i are not small. We thus have a method for the improvement of the theories. This is being used in particular by Eckert on Brown's theory.

80. Problems of the Motion of Other Natural Satellites

The problem of the motion of the Moon is not the only one in the solar system. Very similar situations are encountered in the case of other natural satellites, for example such as Jupiter VI and VII. In the other cases, however, the ratio between the forces is different, or else, two satellites of the same celestial body interact with each other.

Certain satellites, notably Jupiter VIII, represent an extreme case of the theory of the Moon. In this three-body problem the parameters n'/n, e, and i are no more of the order of 0.1 or 0.05 as for the Moon, but are respectively equal to 0.17, 0.4, and 0.5.

The convergence of the expansions is much less satisfactory than in the case of the Moon. The osculating eccentricity sometimes even exceeds the value of 0.67, beyond which the series in the two-body problem no longer converge (cf. section 44).

However, approximate solutions can be obtained, with the aid of methods which differ greatly from the preceding ones. They are based more on purely numerical rather than on dynamic considerations. The approximate solutions are then improved by successive approximations similar to those outlined in the previous section.

In other cases, two satellites of the same body exhibit commensurable mean motions. This is the case with Triton and Hyperion (satellites of Saturn), for which the ratio between the mean motions is 3 : 4. Let l_1 and l_2 be the mean anomalies of the two satellites; the period of $3l_1 - 4l_2$ will then be practically zero. As a result of the small divisor, the contribution of terms with this argument will be increased in the integration. The problem is exactly analogous with that of an artificial satellite in the vicinity of critical inclination. Resonance takes place between the two bodies, and the motion of Hyperion, having the smaller mass, is characterized by libration.

The same resonance and libration are found with the first three satellites of Jupiter, whose mean motions are in a ratio of 1:2:4. The resonance is such that these satellites cannot assume just any configuration: if two of them are in conjunction, the third one is in opposition.

The fourth satellite is not in resonance but is perturbed by the Sun and by the first three satellites. However, the six-body problem is very complicated, and no method has yet given a satisfactory solution of this case.

CHAPTER VII

THE PLANETARY THEORY

The study of planetary motion differs greatly from that of the motion of satellites. Although the great distances of the planets does not allow us to expect the absolute accuracy obtained with the Moon, the difficulty is such that all the theories so far proposed for planetary motions are less satisfactory than the theory of the Moon. The perturbations exerted by planets P_i on another planet P are expanded in terms of the masses of the perturbing bodies. We can perform the calculation of the first-order terms which refer to the perturbations P would experience if the perturbing bodies P_i were moving in Keplerian orbits. However, the calculation of the second-order and higher terms is very complicated. Less theoretical work has been done in this field than in connection with the Moon, and most of the works are of a numerical nature. The outstanding names in this connection are Laplace, Hansen, Le Verrier, and Newcomb.

In this Chapter, we shall follow this general trend and give a numerical rather than an analytical treatment of the problem.

81. The Disturbing Function

As we have seen in section 31, the disturbing function in the three-body problem (Sun, perturbed planet P, and perturbing planet P') is given by

$$R = km' \left(\frac{1}{\varDelta} - \frac{x'x + y'y + z'z}{r'^3} \right) \tag{1}$$

where m' is the mass of the perturbing planet with heliocentric coordinates x', y', and z' $(r' = \sqrt{x'^2 + y'^2 + z'^2})$, x, y, and z are the coordinates of P, and $\varDelta$ (the distance between P and P') is given by

$$\varDelta^2 = (x - x')^2 + (y - y')^2 + (z - z')^2$$
$$= r'^2 + r^2 - 2rr' \cos S$$

S being the angle between the radius vectors of the two planets.

The first significant difference between this case and the theory of the Moon lies in the properties of this disturbing function R, which can be written as a function of S,

$$R = km' \left(\frac{1}{\sqrt{r^2 + r'^2 - 2rr' \cos S}} - \frac{rr' \cos S}{r'^3} \right) \tag{2}$$

We must distinguish from now on between the case where $r > r'$ and the case where $r' > r$, but the two cases are treated in a similar manner. We shall assume that $r' > r$, this being the case of the asteroids perturbed by Jupiter, or the case of Jupiter perturbed by Saturn.

We can now write that

$$R = \frac{km'}{r'}\left[\frac{1}{\sqrt{1 + \left(\frac{r}{r'}\right)^2 - 2\frac{r}{r'}\cos S}} - \left(\frac{r}{r'}\right)\cos S\right] \qquad (3)$$

This expression is formally identical with that for R in the case of the Moon (cf. section 45).

However, the value of r/r' is much too large to be considered as infinitely small. It can reach 0.8 and is equal to 1 in the case of Jupiter and the Trojan asteroids. If we were to carry out the expansion given in section 45, we would have to retain a very great number of terms. The simplified form:

$$R_1 = \frac{km'}{r'}\frac{r^2}{r'^2}\left(-\tfrac{1}{2} + \tfrac{3}{2}\cos^2 S\right)$$

would give no valid indication even as to the qualitative form of the perturbations.

Note that when r/r' is not small, the smallness of R is guaranteed by the fact that km' is small in comparison with kM (M = mass of the Sun), the coefficient of the principal potential term kM/r.

82. First-Order Solution

The second difference from the theory of the Moon partly compensates for the difficulties due to the large value of r/r'. Thus, unlike the case of satellites, in the case of the planets, the secular motions of the nodes and the perhelions are very slow. Thus, the period of the revolution of the nodes is 18 years and that of the perigee is 9 years in the case of the Moon, while the corresponding periods for Mars are respectively 46000 and 20000 years. In fact, the values always exceed 18000 years for all the principal planets.

Since the equations and the disturbing function are identical with those used in the theory of the Moon, the solution will also be in the same form. Thus, in *the first-order theory* (the equivalent of the principal problem of the Moon), in which the perturbing planet is assumed to be in a fixed elliptical orbit, the solution depends on the same four arguments linear in t, i.e. the mean motions l and l' of the two planets and the mean motions Ω and $\bar\omega$ of the nodes and the perihelion in the form of e.g.

$$\Sigma A_{\alpha\beta\gamma\delta}\cos\left(\alpha l + \beta\bar\omega + \gamma\Omega + \delta l'\right) \qquad (4)$$

where α, β, γ, and δ are any four integers.

However, since the periods of $\bar{\omega}$ and $\bar{\Omega}$ are very long, we can write that

$$
\begin{aligned}
\bar{\omega} &= \omega_0 + \omega_0' t \\
\bar{\Omega} &= \Omega_0 + \Omega_0' t
\end{aligned}
\tag{5}
$$

where ω_0' and Ω_0' are very small. Since the theory is required to hold only for a few hundred years, $\beta\omega_0' + \gamma\Omega_0'$ is small. We can then write (4) in the following form:

$$
\Sigma A_{\alpha\beta\gamma\delta} [\cos(\alpha l + \delta l') \cos(\beta\bar{\omega} + \gamma\bar{\Omega}) - \sin(\alpha l + \delta l') \sin(\beta\bar{\omega} + \gamma\bar{\Omega})]
\tag{6}
$$

and expand $\cos(\beta\bar{\omega} + \gamma\bar{\Omega})$ and $\sin(\beta\bar{\omega} + \gamma\bar{\Omega})$ for example in Taylor series of t:

$$
\begin{aligned}
\cos(\beta\bar{\omega} + \gamma\bar{\Omega}) &= \cos[\beta\omega_0 + \gamma\Omega_0 + (\beta\omega_0' + \gamma\Omega_0')t] \\
&= \cos(\beta\omega_0 + \gamma\Omega_0) - (\beta\omega_0' + \gamma\Omega_0')t \sin(\beta\omega_0 + \gamma\Omega_0) + \cdots
\end{aligned}
$$

In the same way

$$
\sin(\beta\bar{\omega} + \gamma\bar{\Omega}) = \sin(\beta\omega_0 + \gamma\Omega_0) + (\beta\omega_0' + \gamma\Omega_0')t \cos(\beta\omega_0 + \gamma\Omega_0) + \cdots
$$

where $\sin(\beta\omega_0 + \gamma\Omega_0)$ and $\cos(\beta\omega_0 + \gamma\Omega_0)$ are constant; we generally use their numerical values. Eq. (6) is then written as

$$
\Sigma m' C_{\alpha\delta} \cos(\alpha l + \delta l') + \Sigma m' S_{\alpha\delta} \sin(\alpha l + \delta l')
\tag{7}
$$

where $C_{\alpha\delta}$ and $S_{\alpha\delta}$ are polynomials in t. Since m' is a factor in the expression of R, it is also a factor here. Only the terms pertaining to the two-body problem lack m' as a factor.

83. Expansion of the Disturbing Function by Harmonic Analysis

The task is then to write the solution in the form of (7), for which reason it is useful to write also the equations in an analogous form. It is not necessary to expand the disturbing function in a Fourier series of all four arguments of l, l', ω, and Ω; it is sufficient to do this only with l and l'.

With the exception of that of Le Verrier's, the classical methods used in the theory of the planets do not require a calculation of the coefficients of the expansion of the disturbing function in an algebraic form. The purely analytical expansion is similar to that given in connection with the Moon, and will not be discussed here. Instead, we shall briefly describe a method which permits a numerical expansion of a'/Δ (cf. section 81). Any other function of the same type is expanded similarly.

We begin by writing, with undeterminate coefficients

$$
\begin{aligned}
F(l, l') = \frac{\Delta^2}{a'^2} &= 1 + \sum_{i=0}^{N} \sum_{j=-N'}^{N'} A_{ij} \cos(il + jl') + \sum_{i=0}^{N} \sum_{j=-N'}^{N'} B_{ij} \sin(il + jl') \\
&= 1 + \sum_{i=0}^{N} \sum_{j=0}^{N'} [(A_{ij} + A_{i-j}) \cos il \cos jl' + (-A_{ij} + A_{i-j}) \sin il \sin jl' \\
&\quad + (B_{ij} + B_{i-j}) \sin il \cos il' + (B_{ij} - B_{i-j}) \cos il \sin jl']
\end{aligned}
\tag{8}
$$

where N and N' are integers chosen *a priori*, such that the following terms can be assumed to be negligible. Furthermore, $A_{i-0}=B_{i-0}=0$.

The formulae used in the two body problem can be employed to calculate the following quantities for any numerical values of l and l':

$$\frac{r}{a}, \quad \frac{r'}{a'} \quad \text{and} \quad \cos S = \frac{xx' + yy' + zz'}{rr'}$$

so that we can have as many values as we wish for the function $F(l, l')$.

Note that if we calculate, for example

$$F(l, l') + F(2\pi - l, l')$$

we obtain

$$2 + \sum_{i=0}^{N} \sum_{j=0}^{N'} 2\left[(A_{ij} + A_{i-j}) \cos il \cos jl' + (B_{ij} - B_{i-j}) \cos il \sin jl'\right]$$

the other two terms having been cancelled out.

Repeating this process several times, we find that

$$F(l, l') + F(2\pi - l, l') + F(l, 2\pi - l') + F(2\pi - l, 2\pi - l')$$
$$= 4 + \sum_{i=0}^{N} \sum_{j=0}^{N'} (A_{ij} + A_{i-j}) \cos il \cos jl'$$

$$F(l, l') - F(2\pi - l, l') - F(l, 2\pi - l') + F(2\pi - l, 2\pi - l')$$
$$= \sum_{i=0}^{N} \sum_{j=0}^{N'} 4(-A_{ij} + A_{i-j}) \sin il \sin jl'$$

$$F(l, l') - F(2\pi - l, l') + F(l, 2\pi - l') - F(2\pi - l, 2\pi - l')$$
$$= \sum_{i=0}^{N} \sum_{j=0}^{N'} 4(B_{ij} + B_{i-j}) \sin il \cos jl'$$

$$F(l, l') + F(2\pi - l, l') - F(l, 2\pi - l') - F(2\pi - l, 2\pi - l')$$
$$= \sum_{i=0}^{N} \sum_{j=0}^{N'} 4(B_{ij} - B_{i-j}) \cos il \sin jl'$$

$$(9)$$

We are thus confronted with the analysis of a function of the form

$$\Phi(l, l') = \sum_{i=0}^{N} \sum_{j=0}^{N'} C_{ij} \cos il \cos jl' \qquad (10)$$

or similar functions with sine instead of cosine terms, these functions being treated in an analogous manner.

In the analysis of $\Phi(l, l')$, for example, we give l' a fixed value l'_0. We are thus considering a one-variable function $\Phi(l, l'_0)$, in which l can take any value.

Assume that the expansion is limited at the $A_3 \cos 3l$ term, i.e.

$$\Phi(l, l'_0) = A_0 + A_1 \cos l + A_2 \cos 2l + A_3 \cos 3l$$

it is sufficient to give l the values 0, $\pi/4$, $\pi/2$ and π to obtain four equations with four unknowns, which can then be solved.

We can calculate similarly the coefficients of the more general expansion:

$$\Phi(l, l_0') = \sum_{i=0}^{N} A_i(l_0') \cos il \tag{11}$$

We can then give l_0' any numerical value and carry out the previous calculation, using N' special values for each coefficient $A_i(l_0')$. These values are so chosen as to be amenable to a calculation similar to that with $\Phi(l, l_0')$, leading to $N+1$ series:

$$A_i(l') = \sum_{j=0}^{N'} C_{ij} \cos jl'$$

where the C_{ij} are constants that no longer depend on any parameter. Combining this expression with (11), we obtain

$$\Phi(l, l') = \sum_{i=0}^{N} \sum_{j=0}^{N'} C_{ij} \cos il \cos jl'$$

Identification with (10) shows that C_{ij} does indeed have the same meaning in both cases.

The same calculation can be performed on any of the four functions defined by (9). Doing this, we arrive at the four quantities:

$$A_{ij} + A_{i-j}; \quad -A_{ij} + A_{i-j}$$
$$B_{ij} + B_{i-j}; \quad B_{ij} - B_{i-j}$$

and thus at A_{ij}, A_{i-j}, B_{ij}, and B_{i-j}. We therefore obtain the required expansion of $F(l, l')$ in the form of (8).

NOTE:

The rules formulated by Brown enable us to give l_0 and l_0' particular values minimizing the errors.

We can also expand in the form of (8) not only $(\Delta/a')^2$, but any known function of the elliptical elements of the two bodies. Furthermore, if we know one, we can find and refine the others by operations on series. Let us assume for example that we know an expansion of $(\Delta/a')^2$ which converges rapidly and for which we need relatively few terms. Suppose further that we have also calculated in the same manner an approximate expansion of a'/Δ. Let us show how to improve it.

We shall denote by S the approximate series of a'/Δ, and by T the exact expansion of $(\Delta/a')^2$. Furthermore, ΔS will denote the correction for improving S. We require that

$$T = \frac{1}{(S + \Delta S)^2}$$

If we neglect ΔS^2, we have that

$$T = \frac{1}{S^2 + 2S\,\Delta S}$$

or

$$S^2 T = \frac{1}{1 + \dfrac{2\,\Delta S}{S}} \# 1 - \frac{2\,\Delta S}{S}$$

and hence

$$\Delta S = \tfrac{1}{2}(S - S^3 T)$$

We obtain ΔS by successive multiplication of the Fourier series S^2, S^3, and $S^3 T$, followed by termwise subtraction.

We can then put that $S_1 = S + \Delta S$ and repeat the procedure with S_1 until ΔS becomes negligibly small.

84. Other Numerical Expansions

Other methods for the expansion of the disturbing function in the form of (8) are based on the Fourier expansion of $(1 - 2\alpha \cos\theta + \alpha^2)^{-s/2}$ where s is an integer and $\alpha < 1$:

$$(1 - 2\alpha \cos\theta + \alpha^2)^{-s/2} = \tfrac{1}{2}b_{s/2}^0 + \sum_{j=1}^{\infty} b_{s/2}^{(j)} \cos j\theta$$

The numbers $b_{s/2}^{(j)}$ are called Laplace coefficients and have been studied and tabulated.

The expansion with Laplace coefficients is based on the facts that r/r' is close to $a/a' = \alpha$, and that the expansion of

$$\left[1 - 2\frac{r}{r'}\cos S + \left(\frac{r}{r'}\right)^2 \right]^{-s/2}$$

is close to that of

$$[1 - 2\alpha \cos S + \alpha^2]^{-s/2}$$

If need be, analytical methods similar to those used in connection with satellites may be used, despite their slow convergence. We can put these expressions in the required form by substitution of the mean values of the elements.

85. Perturbations by Forces in Rectangular Coordinates

The method that is most often used in the construction of a planetary theory is that proposed by Hansen for the Moon, which has been simplified in the case when the motion of the perifocus is not significant.

Omitting the details, which would require elaborate proofs, we shall concentrate on the main characteristics of the method.

The method stars with approximate elliptical motion on an auxiliary ellipse of fixed shape and dimensions, placed in the plane of the osculating orbit and having a perigee with a specified forward motion. Unlike the case of the Moon, we are not considering here the osculating ellipse as an intermediate orbit, this role being now played by an arbitrary fixed ellipse. The perifocus is considered to be fixed in the planetary theory, whilst in the theory of the Moon, we ascribed to it a rotation coinciding with the observed mean motion. The perturbations are described as relative variations of the radius vector and variations of the mean longitude of a body moving in the auxiliary ellipse in accordance with the formulae of the two-body problem.

The disturbing force is first resolved into two components: a radial component R a component perpendicular to the radius vector in the orbital plane, S and a component which is perpendicular to the plane of the orbit W. Using Lagrange equations, one can show that they are equivalent to

$$
\left.
\begin{aligned}
\frac{da}{dt} &= \frac{2e \sin v}{n\sqrt{1-e^2}} R + \frac{2a\sqrt{1-e^2}}{nr} S \\[2ex]
\frac{de}{dt} &= \frac{\sqrt{1-e^2}\sin v}{na} R + \frac{\sqrt{1-e^2}}{na^2 e}\left[\frac{a^2(1-e^2)}{r} - r\right] S \\[2ex]
\frac{di}{dt} &= \frac{r \cos(\omega + v)}{na^2\sqrt{1-e^2}} W \\[2ex]
\frac{d\Omega}{dt} &= \frac{r \sin(\omega + v)}{na^2\sqrt{1-e^2}\sin i} W \\[2ex]
\frac{d\omega}{dt} &= \frac{-\sqrt{1-e^2}\cos v}{nae} R + \frac{\sqrt{1-e^2}\sin v}{nae}\left(1 + \frac{r}{a(1-e^2)}\right) S \\[2ex]
&\quad - \frac{r\sin(\omega+v)\cotg i}{na^2\sqrt{1-e^2}} W \\[2ex]
\frac{dM}{dt} &= n - \frac{1}{na}\left(2\frac{r}{a} - \frac{(1-e^2)\cos v}{e}\right) R - \frac{(1-e^2)\sin v}{nae}\left(1 + \frac{r}{a(1-e^2)}\right) S
\end{aligned}
\right\}
\tag{12}
$$

where n is obtained from a, calculated by the first equation.

NOTE:

This system enables us to study the perturbations caused by any system of forces. It is particularly suitable in the case of non-gravitational forces such as atmospheric friction and radiation pressure experienced for example by artificial satellites. The calculation is carried out as follows: we denote the disturbing function by $\mathscr{R}$ and the components along the three fixed rectangular axes Ox, Oy, and Oz by

$$
\left.
\begin{aligned}
m\frac{\partial \mathcal{R}}{\partial x} &= R[\cos(\omega + v)\cos\Omega - \sin(\omega + v)\sin\Omega\cos i] \\
&\quad - S[\sin(\omega + v)\cos\Omega + \cos(\omega + v)\sin\Omega\cos i] + W\sin\Omega\sin i \\
m\frac{\partial \mathcal{R}}{\partial y} &= R[\cos(\omega + v)\sin\Omega + \sin(\omega + v)\cos\Omega\cos i] \\
&\quad - S[\sin(\omega + v)\sin\Omega - \cos(\omega + v)\cos\Omega\cos i] - W\cos\Omega\sin i \\
m\frac{\partial \mathcal{R}}{\partial z} &= R\sin(\omega + v)\sin i + S\cos(\omega + v)\sin i + W\cos i
\end{aligned}
\right\} \quad (13)
$$

Using the expressions for x, y, and z as a function of the elliptical elements [eq. (25) in section 15], we can write that

$$
\left.
\begin{aligned}
\frac{\partial \mathcal{R}}{\partial a} &= \frac{\partial \mathcal{R}}{\partial x}\frac{\partial x}{\partial a} + \frac{\partial \mathcal{R}}{\partial y}\frac{\partial y}{\partial a} + \frac{\partial \mathcal{R}}{\partial z}\frac{\partial z}{\partial a} \\
\frac{\partial \mathcal{R}}{\partial e} &= \frac{\partial \mathcal{R}}{\partial x}\frac{\partial x}{\partial e} + \frac{\partial \mathcal{R}}{\partial y}\frac{\partial y}{\partial e} + \frac{\partial \mathcal{R}}{\partial z}\frac{\partial z}{\partial e}
\end{aligned}
\right\} \quad (14)
$$

the analogous expressions being obtained for i, Ω, ω, and M.

System (12) is obtained by substituting (13) into (14), and then substituting the result into the Lagrange equations.

86. Variables in Hansen's Method

To arrive at Hansen's method, we can simplify (12) to a system giving the perturbations of the radius vector and the mean longitude as a function of a single quantity $\bar{W}$.

Let us consider an auxiliary ellipse with a semi-major axis a_0, eccentricity e_0, and mean motion n_0. We have that $\mu = n_0^2 a_0^3$.

The independent variable z is such that the mean anomaly of the moving point is always $n_0 z$. Furthermore, the classical relationships (see section 14) pertaining to the coordinates of the moving point M_0 are retained:

$$
\left.
\begin{aligned}
r_0\cos v_0 &= a_0(\cos E_0 - e_0) \\
r_0\sin v_0 &= a_0\sqrt{1 - e_0^2}\sin E_0 \\
n_0 z &= E_0 - e_0\sin E_0
\end{aligned}
\right\} \quad (15)
$$

where r_0, v_0, and E_0 are respectively the radius vector, the true anomaly, and the eccentric anomaly on the ellipse. Denoting by P_0 a fixed point from which we can measure the longitudes in the auxiliary plane, we find that the true longitude is given by $\widehat{P_0 M_0} = \bar{\omega}_0 + gt + v_0$. We choose P_0 in such a way that this longitude is equal to the true longitude of the perturbed body.

NOTE:

In the above expression, gt represents the mean motion of the perifocus, retained in the theory of the Moon but neglected in the theory of the planets, as it will be in the following discussion.

Since we have taken care that the ellipse should be in the osculating plane, point M marking the perturbed body is also in the same plane. Its true longitude is

$$l = v + \chi$$

where χ is the longitude of the osculating perihelion measured from P_0. We have that

$$l = v + \chi = \varpi + v \tag{16}$$

We can *choose z at any moment in the auxiliary ellipse in such a manner that the radius vector of the auxiliary ellipse coincides with that of the perturbed planet.* As a result of such a choice, z is not a linear function of time t. In fact, the variations of z with t describe the perturbations in the direction of the radius vector (and thus of the true longitude) as a function of t. Knowing the function $z(t)$, we can use (15) to find the function $v_0(t)$, and then with (16), the true longitude of the perturbed body:

$$l = v_0(t) + \varpi_0$$

Having defined the radius vector, we obtain the position of the corresponding point of the auxiliary ellipse from

$$r_0(t) = a_0(1 - e_0 \cos E_0)$$

whilst the distance to the origin of the perturbed body will be denoted by r.

We denote by v the quantity

$$v = \frac{r}{r_0} - 1$$

If we have the variations $v(t)$ as a function of t, then

$$r(t) = r_0(t)(1 + v(t))$$

The position of the perturbed body in the plane of the osculating ellipse is thus defined by the functions $z(t)$ and $v(t)$.

Finally, the variation in the position of the osculating plane is defined by the latitude difference $\delta\beta$ between an imaginary moving body (in a plane of fixed inclination, with the same longitude and radius vector) and the actual moving body in the osculating plane.

The combination of $z(t)$, $v(t)$, and $\delta\beta(t)$ defines completely the position of a body at any moment t. We shall give only the equations in z and v, defining the variations in the plane.

87. Calculation by Hansen's Method

Starting with (12), we can find the differential equations in z, v, and β, but the calculations are rather long. The resulting equations are of the first order, and their right-hand sides can be expressed in terms of R, S, and W, dz/dt and dv/dt and are of the form:

$$\frac{dz}{dt} = 1 + W + \frac{a_0 \sqrt{1 - e^2}}{a \sqrt{1 - e_0^2}} \left(\frac{v}{v + 1}\right)^2$$

$$\frac{dv}{dt} = -\frac{1}{2} \frac{\partial W}{\partial z}$$

The expression for $\partial W/\partial t$, in which z is treated as constant (with the aid of r_0 and v_0), can be calculated as a function of R, S, and W, and can then be expanded – as shown at the beginning of this Chapter for the disturbing function – in the following form:

$$\frac{\partial W}{\partial t} = \sum_{i,j,k} C_{ijk} \cos(il + jl' + k\lambda) + \sum_{i,j,k} S_{ijk} \sin(il + jl' + k\lambda)$$

where λ is the mean anomaly of the auxiliary ellipse (and thus a function of z), and l and l' are the mean anomalies of the planet and of the perturbing body, as a function of t.

Some of the terms will be of the form

$$C_{00k} \cos k\lambda \qquad \text{and} \qquad S_{00k} \sin k\lambda,$$

which are constant with respect to t. Integration will thus introduce terms in the form of $t \cos k\lambda$ and $t \sin k\lambda$.

We can replace λ by l (since $\lambda = l$), so that

$$W = \sum_{ij} C_{ij} \cos(il + jl') + \sum_{ij} S_{ij} \sin(il + jl') + \sum_i tC_i \cos il + \sum_i tS_i \sin il$$

The same applies to the form of v and z, obtained from W by integration, which therefore also contain terms in $t^2 \sin il$ and $t^2 \cos il$.

These secular or mixed terms do indeed have the properties envisaged in section 82 [eq. (7)]. The integration constants must be determined by comparing the theoretical results with observations.

Finally, a similar treatment involving a new function U (different from W) enables us to find the terms of the latitude.

88. Higher-Order Planetary Theories

We have discussed briefly one of the possible methods for constructing a theory of the motion of a planet P perturbed by another planet P_i moving in a Keplerian orbit. However, there are other methods, such as those of Le Verrier or Newcomb, each of which has its own advantages and drawbacks. Although they use different variables and parameters, they all lead to expressions of the type of (7).

The complete first-order theory of the motion of a planet P will be the sum of theories obtained by considering separately the planets P_i. In accordance with (7) the general form of the solution is

$$\sum_{i,j,k} m_i C_{ijk}(t) \cos(jl + kl'_i) + \sum_{ijk} m_i S_{ijk}(t) \sin(jl + kl'_i) \tag{17}$$

where m'_i and l'_i are the mass and the mean anomaly of the planet P_i.

However, the effect of a perturbing planet differs slightly according to whether the planet is considered to move in an elliptical orbit or in a perturbed ellipse. We shall now consider a first-order theory of the perturbing body in the form of (17), and calculate the perturbations experienced by P, the first-order theory of this problem being already known. We can carry out the same operations as in the case of a first-order theory, but we substitute in the disturbing function expressions of the type of (17), instead of expressions derived from the study of elliptical motion. The disturbing function will contain terms having as a factor a product of the masses ($m_i m'_i$ or m_i^2), and will be of the form of

$$\sum_{hijkg} m_h m_i C_{hijkg}(t) \cos(jl + kl'_i + gl'_h) + \sum_{hijkg} m_h m_i S_{hijkg}(t) \cos(jl + kl'_i + gl'_h) \tag{18}$$

Integration by a method which can be correlated with a first-order method (e.g. Hansen's method) leads not only to first-order terms, but also to terms of the type of (18) whose factor comprises two masses and whose trigonometric lines may depend on three different anomalies.

Thus, to extract all the second-order terms, we must examine all the possible pairs formed by the perturbed planet P on the one hand and each of the perturbing planets P_i on the other. This must naturally follow the construction of a first-order theory for all the planets P_i.

In practice, we must first evaluate the perturbations prior to calculating some part of the theory, just as we had to evaluate the error introduced by arresting the series expansion of the disturbing function or analogous quantities. At the accuracy of modern observations, the calculations must be arrested only at a fairly late stage.

Thus, Clemence's theory of Mars, based on Hansen's method, includes some third-order terms, in particular those depending on the cubed mass of Jupiter and the squared mass of Jupiter multiplied by the mass of Saturn. Altogether, nearly 1000 arguments have been retained for the final theory.

Some third-order terms also appear in Hill's theory (1890) of Jupiter and Saturn. Most other theories do not go beyond first-order terms.

89. Purely Numerical Methods

We have examined theories increasing in complexity from the case of artificial satellites in a simple gravitational field, via the case of the Moon, up to the complete theory of a planet such as Mars. The greater the complexity of a problem in celestial mechanics, the more we are forced to abandon analytical and algebraic expansions in favour of expansions in which one or more parameters are purely numerical, leading to purely numerical expressions, as in Hansen's theory.

A further change lies in not giving the solution in the form of a Fourier series. We can devise methods of numerical integration giving only the ephemerides of the celestial body in question, i.e. the positions at a number of instants separated by a fixed time interval, the position at any intermediate instant being obtained by simple interpolation. As we shall see, these methods are no longer algebraic or analytical, and they call only for a numerical analysis independent of the concepts of mechanics and the theory of differential equations. These methods lend themselves well to calculation by computers and are very often used, particularly in the case of the planets. We shall now describe Cowell's method, which is used most often, and will then discuss the importance of methods of this type.

90. The Form of Numerical Integration

Consider a function $f(t)$ whose values at regular time intervals h are known. These time intervals are called steps. Let these times be $t_0, t_0+h, t_0+2h, ..., t_0-h, t_0-2h$, etc., the corresponding values of the function f being $f_0, f_1, f_2, ..., f_{-1}, f_{-2}$, etc.

We shall construct a table of the central differences by calculating the first differences with the formula

$$f^{1}_{\frac{n+(n+1)}{2}} = f_{n+1} - f_n$$

TABLE III

arguments	$''f$	$'f$	f	f^1	f^2	f^3	f^4	f^5	f^6
		$'f_{-7/2}$							
$t_0 - 3h$	$''f_{-3}$		f_{-3}						
		$'f_{-5/2}$		$f^1_{-5/2}$					
$t_0 - 2h$	$''f_{-2}$		f_{-2}		f^2_{-2}				
		$'f_{-3/2}$		$f^1_{-3/2}$		$f^3_{-3/2}$			
$t_0 - h$	$''f_{-1}$		f_{-1}		f^2_{-1}		f^4_{-1}		
		$'f_{-1/2}$		$f^1_{-1/2}$		$f^3_{-1/2}$		$f^5_{-1/2}$	
t_0	$''f_0$		f_0		f^2_0		f^4_0		f^6_0
		$'_{1/2}$		$f^1_{1/2}$		$f^3_{1/2}$		$f^5_{1/2}$	
$t_0 + h$	$''f_1$		f_1		f^2_1		f^4_1		
		$'f_{3/2}$		$f^1_{3/2}$		$f^3_{3/2}$			
$t_0 + 2h$	$''f_2$		f_2		f^2_2				
		$'f_{5/2}$		$f^1_{5/2}$					
$t_0 + 3h$	$''f_3$		f_3						
		$'f_{7/2}$							

and then the second differences:

$$\frac{f^2_{n+(n+1)}}{2} = f^1_{n+1} - f^1_n$$

differences of the p-th order being calculated from differences of the order of $p-1$ by

$$\frac{f^p_{n+(n+1)}}{2} = f^{p-1}_{n+1} - f^{p-1}_n \tag{29}$$

We shall complete the table to the left with the aid of the quantities $'f$ and $''f$ such that

$$\left.\begin{array}{l} f_n = 'f_{n+1/2} - 'f_{n-1/2} \\ 'f_n = ''f_{n+1/2} - ''f_{n-1/2} \end{array}\right\} \tag{30}$$

These quantities can be calculated from the previous column only to one additive constant.

We now define the following additional quantities:

$$\left.\begin{array}{l} 'f_n = \frac{1}{2}('f_{n+1/2} + 'f_{n-1/2}) \\ f^p_n = \frac{1}{2}(f^p_{n+1/2} + f^p_{n-1/2}) \end{array}\right\} \tag{31}$$

When $f(t)$ is equated with d^2x/dt^2, we find the equalities:

$$''f_n = \frac{x(t_0 + nh)}{h^2} - \frac{1}{12}f_n + \frac{1}{240}f^2_n - \frac{31}{60480}f^4_n + \frac{289}{3628800}f^6_n + \cdots \tag{32}$$

$$'f_{n\pm1/2} = \frac{x'(t_0 + nh \pm (h/2))}{h} \pm \frac{1}{2}f_n + \frac{1}{12}f^1_n - \frac{11}{720}f^3_n + \frac{191}{60480}f^5_n + \cdots \tag{33}$$

where x' is the value of the derivative of x with respect to t at the indicated instant.

91. Starting the Numerical Integration

We assume that the initial conditions of the motion, i.e. the coordinates x_0, y_0, and z_0 and the velocities x'_0, y'_0, and z'_0 of the body are known. The differential equations are of the form:

$$\left.\begin{array}{l} \dfrac{d^2x}{dt^2} = f(x, y, z, x', y', z', t) \\[2mm] \dfrac{d^2y}{dt^2} = g(x, y, z, x', y', z', t) \\[2mm] \dfrac{d^2z}{dt^2} = h(x, y, z, x', y', z', t) \end{array}\right\} \tag{34}$$

The quantities x', y', and z' (which are generally absent) may be introduced into the equations by taking into account the friction of the medium. The coordinates of the perturbing bodies are assumed to be known as a function of time, and they introduce t into the right-hand sides of the above differential equations. We assume that formulae (32) and (33) are truncated respectively at terms in f^6 and f^5, and proceed in the following manner:

1. We first assume that all the differences $f^1, ..., f^n$ are zero. We can calculate f_0 by the use of x_0, y_0, z_0, and t_0. Using (32) and (33), we then deduce the quantities $"f_0, 'f_{-\frac{1}{2}}$, and $'f_{\frac{1}{2}}$. The same calculation is carried out to find the analogous quantities g and h.

2. We complete the column of the f terms in Table III on the assumption that they are all the same. Using the previous hypotheses, we thus insert $f_{-3}, f_{-2}, f_{-1}, f_1, f_2$, and f_3. We then do the same in tabulating the g and the h terms.

3. We can now calculate from (30) all the $'f$ and $"f$ terms shown in Table III. Formulae (32) and (33) thus enable us to calculate the coordinates and the velocities for each value of t. The results of these calculations are substituted in f in accordance with (34). We can therefore calculate all the elements of the triangle in the table with the aid of (31).

4. A basic hypothesis in this method of numerical integration consists in assuming that all the f^6 terms are equal to f_0^6 which has just been calculated. Using (31) we then construct a rectangular table by calculating all the f_j^i terms such that $i \leq 5$ and $-7/2 \leq j \leq 7/2$.

5. We can calculate $x_{-3}, x_{-2}, x_{-1}, x_1, x_2$, and x_3, as well as the corresponding x' values, by using (32) and (33) (this time in the complete form) and the complete rectangular table.

6. The calculations specified in points 3 to 5 above are repeated to construct the tables for g and h.

7. Having thus obtained the new values of the coordinates x, y, and z, and the derivatives x', y', and z' for all the times between $t - 3h$ and $t + 3h$, we can recalculate the f, g, and h terms corresponding to these times, and replace by these new values the quantities in the f, g, and h columns of the tables.

8. We complete the triangle, repeat the calculation from point 4, and continue until new iteration causes no change in the table.

We thus have seven values of unknowns corresponding to the seven times between $t_0 - 3h$ and $t_0 + 3h$. The seventh differences of the corresponding functions f, g, and h are zero.

92. The Numerical Integration Proper

Once the rectangular table has been constructed and stabilized, we can start the numerical integration proper in which we calculate the values of the unknowns at instants $t_0 + nh$ forming an ascending series.

The table of differences from t_0 to $t_0 + 4h$ is shown in Table IV.

We proceed in the following manner to extend the table by one unit (of the indices) and to finalize the values of x, y, z, and x', y', and z'.

TABLE IV

arguments	$''f$	$'f$	f	f^1	f^2	f^3	f^4	f^5	f^6
t_0	$''f_0$		f_0		f_0^2		f_0^4		f_0^6
		$'f_{1/2}$		$f_{1/2}^1$		$f_{1/2}^3$		$f_{1/2}^5$	
$t_0 + h$	$''f_1$		f_1		f_1^2		f_1^4		f_1^6
		$'f_{3/2}$		$f_{3/2}^1$		$f_{3/2}^3$		$f_{3/2}^5$	
$t_0 + 2h$	$''f_2$		f_2		f_2^2		f_2^4		f_2^6
		$'f_{5/2}$		$f_{5/2}^1$		$f_{5/2}^3$		$f_{5/2}^5$	
$t_0 + 3h$	$''f_3$		f_3		f_3^2		f_3^4		f_3^6
		$'f_{7/2}$		$f_{7/2}^1$		$f_{7/2}^3$		$f_{7/2}^5$	
$t_0 + 4h$									

1. We assume that $f_4^6 = f_3^6$ and calculate the following f values with the aid of (31) and then (30):

$$f_{9/2}^5, f_4^4, f_{9/2}^3, f_4^2, f_{9/2}^1, f_4^0, 'f_{9/2} \quad \text{and} \quad ''f_4$$

The same is then done with the tables for g and $\hbar$.

2. Using (32) and (33), we deduce from the above results the values of x_4, y_4, z_4 and x_4', y_4', and z_4', which enable us to recalculate $f_4, g_4,$ and $\hbar_4$.

3. The new value of f_4 enables us to correct the differences $f_{7/2}'$, and consequently $f_3^2, f_{5/2}^3, f_2^4, f_{3/2}^5,$ and f_1^6 [with the aid of eq. (31)]. The same is done in the case of g and $\hbar$.

4. We put $f_2^6, f_3^6,$ and f_4^6 equal to f_1^6 and correct all the differences calculated from these f^6 terms, as in point 4 in the previous section. The same is done in the case of g and $\hbar$.

5. We use (32) and (33) to calculate the values of x, y, z and x', y', and z' for $t_0 + h, t_0 + 2h, t_0 + 3h,$ and $t_0 + 4h$, which enables us to recalculate the corresponding functions $f, g,$ and $\hbar$.

6. We can now recalculate all the $'f$ and $''f$ functions, together with the differences in the three tables.

7. We repeat the calculation from point 4 until stable values are obtained for $f_1, g_1, \hbar_1, x_1, y_1, z_1, x_1', y_1',$ and z_1'.

These values are thus considered as final, and the calculation in point 1 is repeated with terms having indices increased by one unit. We proceed in this manner until the calculation has been extended to the required point in time.

93. Properties of Numerical Integration

Cowell's method of numerical integration described above admits a great number of variants. In one of these, specially adapted to calculation by computers, the calculation of the successive differences in the progression is replaced by calculation of the linear functions of the quantities f and $''f$.

Three different formulae permit (1) the extrapolation of a new value of x to the subsequent steps; (2) improvement of this value from a known approximate value; and (3) verification of the results when the extrapolated values of several subsequent f functions are known.

The variants also differ in the number of terms retained in (32) and (33). This number naturally depends on the required accuracy and the magnitude of h. Furthermore, an accuracy test on each integration enables us to decide whether to double the steps or to halve them.

In another method of numerical integration, devised by Encke and often used in celestial mechanics, we use not the differential equations of motion, but those of the perturbations of an osculating elliptical motion at a given point in time. From time to time, we carry out modifications enabling us to make a correction for a new osculating elliptical orbit, whenever the perturbations are likely to have modified the old orbit to such an extent that the differences lead to a noticeable variation in the disturbing forces. The approximation in this method is carried out to second-order perturbations.

Encke's method is more advantageous when the perturbations are very small, and becomes very difficult when correction of the orbit is required too frequently.

Errors involved in integrations come from two sources: some are due to the fact that certain terms are neglected in (32) and (33) (truncation error), others are entailed in the approximations made in the course of the operations (rounding-off errors). Formulae (32) and (33) being known, we can find a step in which errors of the first type are negligibly small with respect to the required accuracy. However, errors of the second type increase linearly with the $\frac{3}{2}$-th power of the number of steps.

Since the aim of numerical integration is to find the ephemeris of a celestial body in the time interval Δt, one might be tempted to reduce the truncation error by reducing the step. However, this would lead to an increase in the rounding-off errors.

If, on the other hand, we increase the step, we must proceed further with the expansions (32) and (33), so as to reduce the truncation errors. However, owing to the instability of this method and the necessity of increasing the number of significant figures of the calculation, the expansions cannot be extended indefinitely.

The optimum is reached with some 50 or 100 integration steps per revolution and a formula with up to the eighth differences and ten significant figures.

94. The Use of Numerical Integration

The coordinates and velocity components at the first instant are equivalent to the integration constants in the analytical methods, and, similarly to these integration constants, they cannot be directly determined by observation. Therefore, we proceed in the following manner:

1. We perform a first numerical integration with the approximate values of the initial conditions, compare the results with observed values, and draw up a list of $(o-c)$ differences (observation – calculation) for each time of observation.

2. Carrying out six new numerical integrations, we calculate the variations in

these differences resulting from the introduction of a unit variation in each of the six initial values.

3. We determine the values of the variations that we must introduce in the initial values to minimize the $(o-c)$ differences obtained in point 1. If the result is inadequate, the calculation in point 1 can be repeated with the aid of an evaluation method such as the method of least squares.

This is the best way of representing the observations in a certain time interval and of predicting the positions from the final result of the integrations. The technique is used currently to predict the positions of short-term artificial satellites orbiting for a few days. It is possible to take into account and calculate quickly a great number of perturbing factors such as the flattening of the Earth, atmospheric friction, and radiation pressure.

This method is extensively used for studying the motions of planets, e.g. for the compilation of the current ephemerides of Jupiter, Saturn, Uranus, Neptune, and Pluto.

95. Comparison between Numerical Integration and Analytical Theories

Since the limitations mentioned at the end of section 93 restrict the number of revolutions that can be treated by numerical integration, this technique is difficult to apply to satellites, in the case of which the observations involve an excessively great number of revolutions. It is more suitable in the case of planets and short-life artificial satellites.

Numerical integration represents today the fastest and the most efficient means of calculating the exact trajectory of a celestial body in a given time interval. However, despite all these advantages, the method does not satisfy the requirements of astronomers on two accounts.

In the first place it is restricted to a limited time interval, and gives no indication (not even a qualitative one) as to the motion outside this time interval. In contrast, a general theory with a given accuracy in the same interval still holds with good approximation for an interval ten times as great, and gives significant qualitative indications about the motion well outside this interval, particularly when it is a purely trigonometric general theory without any secular terms.

Secondly, astronomers are particularly interested in distinguishing between cause and effect and in identifying the precise origin of any perturbation experienced by a celestial body. Again, this can be achieved with analytical expressions, whilst in the numerical integration the various perturbations are all lumped together and we obtain no indication as to the form of any of them.

Thus, numerical integration may serve as an intermediate step in the construction of an analytical theory by forming an intermediate orbit through the harmonic analysis of the results of the numerical integration. This method has also been used to verify an analytical theory constructed independently. It appears, therefore, that the method is a powerful numerical tool, comparable to operations with Fourier series and harmonic analysis. Nevertheless, the construction of increasingly more accurate analytical theories remains the central task of celestial mechanics.

BIBLIOGRAPHY

Andoyer, H., *Cours de mécanique céleste*. 2 vols. Gauthier-Villars, Paris, 1923–1926.

Brouwer, D. and Clemence, G. M., *Methods of Celestial Mechanics*. Academic Press, New York, 1961.

Brown, E. W. and Shook, C. A., *Planetary Theory*. Dover reprint, New York, 1964.

Chazy, J., *Mécanique céleste*. Presses Universitaires de France, Paris, 1953.

Danjon, A., *Astronomie générale*. Editions Sennac, 1960.

Moulton, F. R., *An Introduction to Celestial Mechanics*. 13th ed. The Macmillan Co., New York, 1959.

Smart, W. M., *Celestial Mechanics*. Longmans, Green & Co., London, 1953.

Tisserand, F., *Traité de mécanique céleste*. 4 vols. Gauthier-Villars, Paris. Vols. 1–2, rev. ed. 1960–1962; vol. 3–4, 1896.